AQA BIOLOGY
Specification A

A New Introduction to

HUMAN BIOLOGY

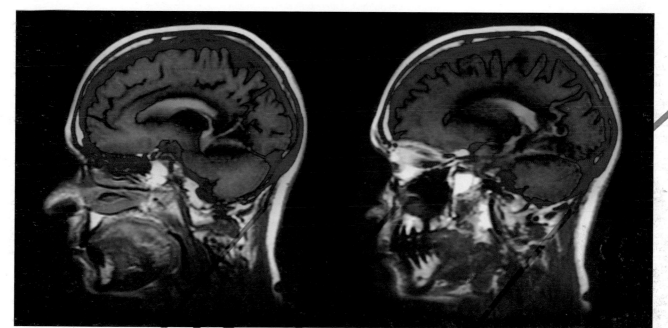

Bill Indge

Martin Rowland

Margaret Baker

Hodder & Stoughton
A MEMBER OF THE HODDER HEADLINE GROUP

Orders: please contact Bookpoint Ltd, 78 Milton Park, Abingdon, Oxon OX14 4TD.
Telephone: (44) 01235 827720, Fax: (44) 01235 400454. Lines are open from 9.00–6.00,
Monday to Saturday, with a 24 hour message answering service. Email address:
orders@bookpoint.co.uk

A catalogue record for this title is available from The British Library

ISBN 0 340 78167 X

First published 2000

Impression number 10 9 8 7 6 5 4 3 2 1
Year   2005 2004 2003 2002 2001 2000

AQA Examination questions reproduced or adapted by permission of the Assessment and
Qualifications Alliance.

Cover photo from: Montage of images from Photodisc.

Photo acknowledgements
Biophoto Associates (Figs 1.3, 1.5, 1.6, 1.7, 1.10, 3.11, 5.18, 5.27, 6.1, 9.1, 9.2, 9.3, 9.4,
9.5, 9.6, 10.4)
Bourne Hall Clinic (Fig. 13.12)
John Birdsall (Fig. 3.10, 4.27, 7.1, 11.2)
Holt Studios International (Figs 3.3, 3.8, 3.30, 4.25, 10.5a, 12.4b, 12.14, 12.15, 12.22a,
12.22b, 12.25, 12.26, 12.29, 13.7b)
Hulton Getty (Fig. 13.15)
Bill Indge (Figs 12.4a, 12.4c, 12.6, 12.16, 12.18, 12.21, 12.22c, 13.9, 13.10)
Natural History Picture Agency (Fig. 3.12, 3.26, 3.36, 7.10, 7.11, 10.5c, 10.10, 13.8)
Planet Earth Pictures (Figs 3.1, 3.32, 4.1, 5.25, 5.26, 6.3, 6.5, 10.5b, 12.1, 12.2, 12.31, 13.1)
Science Photo Library (Figs 1.4, 1.11, 2.1, 2.13, 3.22 3.24, 4.13, 6.10, 6.23, 8.1, 9.8, 9.13,
12.24, 13.2, 13.11, 13.18)
SECOL (Fig. 5.1)
Tropix (Fig. 6.9)
Wellcome Picture Library (Fig. 11.1)

Designed by Allan Sommerville, Cambridge Publishing Management.
Typeset by Cambridge Publishing Management.

Printed in Italy for Hodder & Stoughton Educational, a division of Hodder Headline Plc,
338 Euston Road, London NW1 3BH by Printer Trento.

Contents

Introduction

About AS Human Biology

Human Biology is never far from the headlines. While this book was being written, the human genome has been sequenced and we now know the complete arrangement of the three thousand million bases that make up human DNA. In Africa, one in four Kenyans is now estimated to be HIV-positive and 350 people die every day from AIDS. Smoke is once again blanketing south-east Asia from fires burning out of control in Indoncsia. Biologists are concerned with all these issues. They work in the fields of cell biology, medicine, food production and ecology and the work they do is vital to us all.

This book is an introduction to the study of human biology. It is based on the AQA specification A Biology (Human) AS course. The first four units are concerned with the basic biological principles that underpin nearly all the biology that you will study later. They concentrate on providing you with an understanding of cell biology. You will look at proteins and other biological molecules, enzymes, cell structure and the way in which substances get in and out of cells. All these are topics that you will encounter many times in the rest of your A level course.

Physiology is the study of how living organisms work. In units 5 and 6 you will look at the physiology of the blood and gas-exchange systems. These first six units cover Module 1 of the specification.

Module 2 has as its main theme, human health. This is covered in the remainder of the book. The different units look at various aspects of disease and its control. We start by considering microorganisms and parasites and study the ways in which the immune system and drugs such as antibiotics help us to fight infection. We go on to look at genes, gene technology and genetic disease. The final topics of study are the disease which, in many ways, reflect our modern lifestyle, heart disease and cancer.

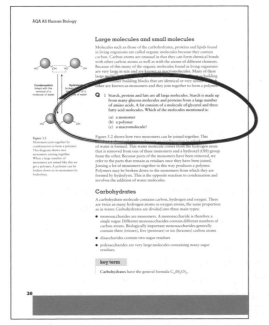

About this book

Before you begin working from this book, it would be a good idea to have a look the way it has been written. Each unit has been presented in the same way and shares a number of features.

The unit opening

We have chosen something topical to start each unit. This is usually an interesting application of biology related to the contents of the unit. It will help to show you how biology plays an important part in the real world.

The text

The most important word in the study of AS Human Biology is "understanding". If you understand the basic ideas, you will always be able to learn the necessary details later. But if you don't understand the underlying biology, you will find it very difficult to remember the facts. In writing this book, we have kept this in mind and have tried to explain the basic ideas as clearly as possible. Most of you will have come to AS Human Biology from a background of GCSE science so we have avoided introducing unnecessary technical terms. Where these terms are essential, they have been explained carefully in the "Key Terms" section. The book has been illustrated in colour throughout. The drawings and the information in the captions should provide you with further help in understanding the basic biology.

Text questions

The text questions should help you to understand what you have just been reading and they are meant to be answered as you go along. Most of them are very straightforward and can be attempted from the information in the paragraph or two which come immediately before. We haven't included any answers. If you get stuck, read the last paragraph again. If you still have problems, make a note of the question and get some help.

Extension boxes

If you restrict yourself to learning the absolute minimum, you should be successful in your AS Human Biology examinations, providing that all you want is a pass. If you want to do better or go on to study the subject at A level, you will need to get a wider understanding of the subject. This is where the extension boxes should help. They are intended to do two things. Not only will they tell you a little more about topics mentioned in the main text, but they will also help you to begin to understand the links between different aspects of biology.

Summary

At the end of each unit is a summary. This tells you everything you ought to know when you have finished the work. It will enable you to check that you have done everything that the specification requires.

Assignment

Studying human biology involves much more than learning facts. It also involves acquiring a range of skills. A biologist should be able to apply facts to new situations; interpret photographs and drawings, graphs and tables, and design experiments. As a biologist, you should also be able to use mathematical skills to carry out a range of simple calculations, and you must be able to use scientific English to communicate your knowledge and ideas effectively. Success in your unit tests will therefore mean that not only will you have understood the basic ideas and learnt the necessary facts, but you will have developed a range of skills. Each unit in this book finishes with an assignment. The purpose of the text in the chapter is to provide you with the factual information you require. The purpose of the assignment is to help you to master the skills you need.

Cells and Cell Structure

key term

Cells
The cell is the basic unit of structure in almost all organisms.

Prokaryotic cells do not have a nucleus.

Eukaryotic cells do have a nucleus.

Soil becomes polluted when land is used for dumping chemical waste. Since March 2000, local authorities in the UK must publish the sites of polluted land. This poses a problem. Because some dumping is done illegally, local authorities do not always know when soil has become polluted. New techniques have been developed to identify land that has been polluted.

One of these new techniques uses cells from earthworms. Earthworms live in soil. They eat soil and digest the organic matter that is in it. Unfortunately, they also absorb many of the harmful substances from polluted soil. These substances are destroyed in the cells of an earthworm's body. Small structures within cells, called lysosomes, are involved in this destruction.

In the new technique, a small sample of cells is taken from earthworms. Done properly, this does not harm the earthworms. These cells are then injected with a red dye. Healthy cells are not stained by the red dye because it is removed by the lysosomes. If an earthworm has passed through polluted soil, its lysosomes are not able to remove the injected dye. The dye spills into the rest of the cell, which then turns red.

Scientists can dig up worms at sites to be tested for pollution, remove a few cells from the worms and inject them with red dye. Cells that quickly turn red show that a worm has been burrowing through polluted soil.

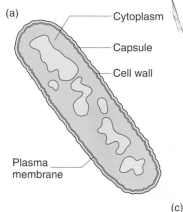

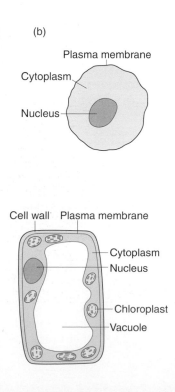

Figure 1.1
(a) A bacterial cell; (b) an animal cell; (c) a plant cell. The bacterial cell does not have a nucleus: it is called a prokaryotic cell. The animal and plant cells, although different from each other, both have a nucleus. They are called eukaryotic cells. (Cells are not to scale)

Two types of cell

Living things are called organisms. Almost without exception, they are made of small units called cells. Some organisms are made from only one cell. Bacteria, yeast and the malarial parasite *Plasmodium* are examples of single-celled organisms. Most organisms are made from many cells. From a single fertilised egg cell, your body grew to about one trillion cells by the time you were born.

Figure 1.1 shows three types of cell: a bacterial cell, an animal cell and a plant cell. These cells differ from each other in a number of ways. One of these is crucial: bacterial cells have no nucleus. Because of this, bacterial cells are called **prokaryotic cells** (*pro* means 'before' and *karyon* means 'nucleus'). Animal and plant cells do have a nucleus. They are called **eukaryotic cells** (*eu* means 'true' and *karyon* means 'nucleus').

The main features shown in each of the cells in Figure 1.1 are summarised in Table 1.1.

Feature	Function	Bacterial cell	Animal cell	Plant cell
Capsule	Protects cell from attack, e.g. by human granulocytes (see Unit 6)	Present	Absent	Absent
Cell wall	Provides strength and stops cell bursting in dilute solutions	Present (murein)	Absent	Present (cellulose)
Plasma membrane	Barrier between cell and its surroundings. Controls movement of substances in to, and out of, cell	Present	Present	Present
Cytoplasm	Performs most of the 'work' of the cell	Present	Present	Present
Nucleus	Contains the genetic code that controls the cell's activities	Absent	Present	Present

Table 1.1 The main features of bacterial cells, animal cells and plant cells. Bacterial cells are prokaryotic. Animal and plant cells are eukaryotic.

Q 1 (a) Name *one* way in which a plant cell is similar to a prokaryotic cell but different from an animal cell.

(b) Name *one* way in which a plant cell is similar to an animal cell but different from a prokaryotic cell.

The optical microscope

You could not see the detail shown in Figure 1.1 with your naked eye. You would need to use an optical microscope to see the detail of animal and plant cells. Even with an optical microscope, you would see only the shape of a bacterial cell.

You will use an optical microscope during your course of study. Like any other skill, your ability to use an optical microscope will improve with practice. Figure 1.2 shows the sort of optical microscope you are likely to use. It has three groups of lenses: the condenser, the objective and the eyepiece. These magnify the object placed on the stage of the microscope. We can control the magnification using different objective lenses. The coarse and fine focus knobs enable us to focus the microscope.

Preparing temporary mounts

Figure 1.3 shows a thin film of plant cells. This film of cells is easy to obtain. To view such a film of cells with an optical microscope, you need to put it on a glass slide. This is called a mount. Because this mount quickly dries up, it does not last for more than a few hours. It is therefore

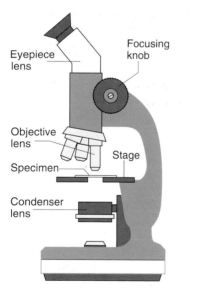

Figure 1.2
An optical microscope. This uses lenses to focus light through a specimen on its stage. This type of microscope can be carried around and is relatively cheap.

Eyepiece lens

Focusing knob

Objective lens

Stage

Specimen

Condenser lens

called a **temporary mount**. Table 1.2 shows the steps you need to follow to prepare a temporary mount in a school or college laboratory.

Figure 1.3
A thin film of cells from the epidermis ('skin') of an onion leaf. A stain has been used to show up the cell walls and the nuclei. Without the stain, we would not be able to see these structures clearly.

Procedure	Explanation
Place a drop of water on a glass slide	
Peel a film of cells from the inner surface of the fleshy leaf of an onion Cut a small piece from this film and float it on the drop of water on the slide	Floating flattens the film of cells. Done properly, this ensures that we see a layer of cells that is only one cell thick 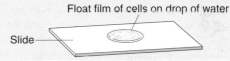
Use a mounted needle to lower a thin cover slip of glass over the layer of cells 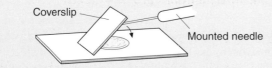	Covers the temporary mount and slows the evaporation of the water
Place a drop of iodine solution at one edge of the cover slip Put a piece of filter paper at the opposite edge of the cover slip Allow the filter paper to soak up the water from under the cover slip This will draw iodine solution under the cover slip	Iodine will stain parts of the cell Some of the cell's structure will now be easier to see 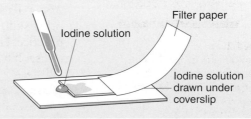
Remove any surplus water or iodine solution from the slide View the slide under the microscope	You will now see cells with cell walls, cytoplasm and nuclei

Table 1.2 Stages in the production of a temporary mount. Many different stains can be used to show cell structure. Iodine is just one of these stains.

The electron microscope

Microscopes are used to make objects look bigger. This is called **magnification** and lets us see details of cell structure that cannot be seen with the naked eye. The optical microscope in Figure 1.2 has a low-power objective lens and a high-power objective lens. A typical low-power objective lens magnifies objects to 10 times their real size. The image is further magnified by the eyepiece lens, which usually magnifies 10 times. Using a 10 × objective lens and a 10 × eyepiece lens means that we see cells 10 × 10, i.e. 100 times, their real size.

Q 2 How much bigger than their real size would we see cells if we used a 10 × eyepiece lens and a 40 × objective lens?

You might expect that a high-power lens would let us see more detail than a low-power lens. Up to a point this is true. However, there comes a point at which increasing the magnification of an optical microscope does not show more detail. To understand this, we need to know something about the physical nature of light.

Light is made from particles (called photons) that travel in a wave. We can only see that two objects are separate from each other if light can pass between them. Light can pass between the letters on this page, so we see the individual letters. When objects are very close together, light cannot pass between them. We cannot see them as separate objects; they appear as a single object. This ability to distinguish between objects is called **resolution**. The size of the wavelength of light limits its **resolution** to about 2 μm.

Electrons have a much smaller wavelength than light. This means that they can pass between objects that are too close together to let light pass between them. In turn, this means that electrons can resolve objects that light cannot resolve. Microscopes that use electrons instead of light are called **electron microscopes**. The resolution of a modern electron microscope is about 1 nm. A **transmission electron microscope** passes beams of electrons through a very thin object. A **scanning electron microscope** bounces beams of electrons off the surface of an object.

Q 3 Explain the difference between magnification and resolution.

Figure 1.4 shows an electron microscope in use. It is much bigger than an optical microscope and so cannot be moved around. Its size is partly caused by:

● the power supply to its electromagnets, which focus beams of electrons

● the pump, which is used to make a vacuum inside its specimen tube.

An electron microscope is also much more expensive than an optical microscope. You are likely to see one only in a university or a scientific research organisation.

Table 1.3 summarises the major differences between optical microscopes and electron microscopes.

key terms

Magnification describes the extent to which an image has been enlarged.

Resolution describes the ability to distinguish between points that are close together. The resolving power of light is about 2 μm (pronounced 'two micrometres'; 1μm is one millionth of a metre, i.e. 10^{-6} m). The resolving power of electrons is about 1 nm (1 nm, pronounced 'one nanometre', is one thousand-millionth of a metre, i.e. 10^{-9} m).

Figure 1.4
A transmission electron microscope. This type of microscope is very expensive and, unlike an optical microscope, cannot be moved. The tube containing the specimen also contains a vacuum to stop electrons being scattered by air molecules.

Feature	Optical microscope	Electron microscope
Radiation used	Beams of light	Beams of electrons
Method of focusing	Glass lenses	Electromagnets
Maximum magnification of specimen being observed	Typical student microscopes magnify up to 400 × the real size of the specimen The maximum magnification of a research microscope is about 1500 ×	About 500 000 × the real size of the specimen
Resolution	About 2 μm	About 1 nm
Presence of vacuum within microscope	Absent Scattering of light by air molecules does not affect image production	Present Scattering of electrons by air molecules interferes with image production
Specimens that can be viewed	Specimens can be alive or dead; If dead, they can be stained Specimens must be relatively thin so that light can pass through them	Specimens are always dead since they cannot survive inside a vacuum **Transmission microscope:** specimens must be extremely thin so that electrons can pass through them **Scanning microscope:** specimens need not be thin since beams of electrons are bounced off their surface

Table 1.3 A comparison of optical and electron microscopes.

Q 4 Why must electron microscopes contain a vacuum?

Ultrastructure of eukaryotic cells

We can see cell structure using an optical microscope. An electron microscope shows us a cell's ultrastructure. Figure 1.5 shows the ultrastructure of a epithelial cell. The photograph was taken using the image from a transmission electron microscope. Such a photograph is called an **electron micrograph**. Figure 1.5 shows us much more about a eukaryotic cell than Figure 1.3 does. For example, in Figure 1.5 we can see:

- the plasma membrane, which separates the cell from its surroundings

- a nuclear envelope, which separates the nucleus from the cytoplasm

- many different structures, called **organelles**, within the cytoplasm.

Figure 1.5
Electron micrograph of a epithelial cell.

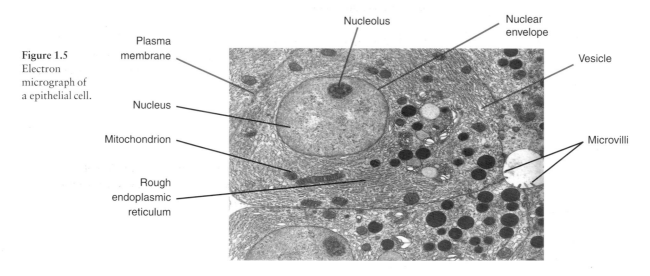

Figure 1.6 is an electron micrograph of a plant cell. It shows many of the same structures you can see in Figure 1.5. It also shows the cell wall and chloroplasts that are not found in animal cells. Table 1.4 summarises the function of the structures shown in Figures 1.5 and 1.6.

Figure 1.6
Electron micrograph of a mesophyll cell.

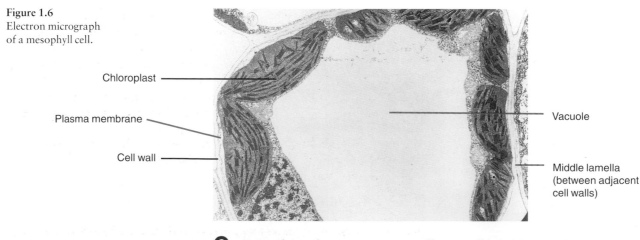

Q 5 Explain why we cannot see cell organelles in Figure 1.3.

Feature	Description	Function
Plasma membrane	Appears as two dark lines These represent two layers of phospholipids (see Chapter 2)	Separates cell from its surroundings Controls movement of substances in to, and out of, cell Allows cell identification and cell adhesion
Adjacent cell walls / Middle lamella	Layers of cellulose fibres around plant cells and of chitin around fungal cells Absent from animal cells	Provides mechanical strength and support Stops cell bursting in dilute solutions
Chloroplast Lamellae Stroma	Cigar-shaped organelle found in photosynthesising cells Its inner membrane has folds, called lamellae, which surround a fluid, called the stroma Chlorophyll is found in the lamellae	Traps light energy and uses it to produce carbohydrates from carbon dioxide and water, i.e. photosynthesis
Endoplasmic reticulum (ER) Ribosome	A series of tubes running through the cytoplasm The tubes are surrounded by membrane Rough ER has ribosomes studded into its membranes; smooth ER has no ribosomes	Speeds up the distribution of substances through the cytoplasm
Golgi apparatus	A stack of flattened sacs surrounded by membrane Small vesicles bud off its edge Not found in plant cells	Stores (and chemically modifies) substances produced in the cell These substances are later secreted using vesicles
Lysosome	A small sphere of liquid surrounded by membrane	The fluid contains powerful protein-digesting enzymes (lysozymes) These organelles are used to digest protein, e.g. they digest the cell after it dies In some white blood cells, they are used to digest bacteria (see Chapter 6)
Microvilli Plasma membrane	Microscopic folds in the plasma membrane Only found in some types of epithelial cells	Increase surface area of membrane for absorption

Table 1.4 The major organelles found in eukaryotic cells (see Figures 1.5 and 1.6).
Continued on next page

Continued from previous page

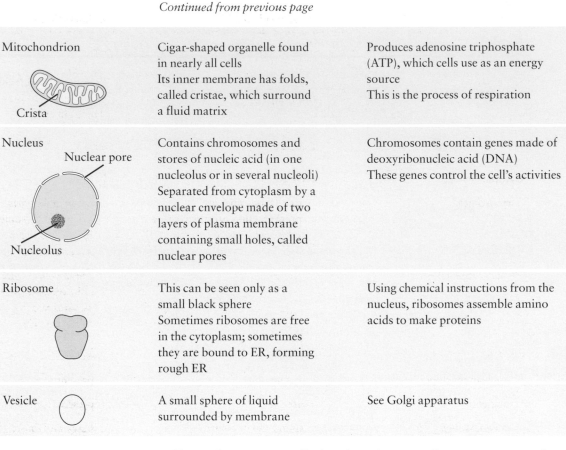

Mitochondrion	Cigar-shaped organelle found in nearly all cells. Its inner membrane has folds, called cristae, which surround a fluid matrix	Produces adenosine triphosphate (ATP), which cells use as an energy source. This is the process of respiration
Nucleus	Contains chromosomes and stores of nucleic acid (in one nucleolus or in several nucleoli). Separated from cytoplasm by a nuclear envelope made of two layers of plasma membrane containing small holes, called nuclear pores	Chromosomes contain genes made of deoxyribonucleic acid (DNA). These genes control the cell's activities
Ribosome	This can be seen only as a small black sphere. Sometimes ribosomes are free in the cytoplasm; sometimes they are bound to ER, forming rough ER	Using chemical instructions from the nucleus, ribosomes assemble amino acids to make proteins
Vesicle	A small sphere of liquid surrounded by membrane	See Golgi apparatus

Table 1.4 The major organelles found in eukaryotic cells (see Figures 1.5 and 1.6). The organelles are not drawn to the same scale.

Ultrastructure of prokaryotic cells

Figure 1.7 is an electron micrograph of a bacterial cell. In it, you can see the capsule, cell wall and cytoplasm shown diagrammatically in Figure 1.1. You can also see the genetic material (light region). Notice in the prokaryotic cell in Figure 1.7 that:

● the genetic material is not surrounded by a nuclear envelope, as it is in the eukaryotic cells shown in Figures 1.5 and 1.6. Instead, the single, circular chromosome is free in a special region of the cytoplasm

● the cytoplasm does not contain most of the cell organelles shown in Figures 1.5 and 1.6. The only organelle present is the ribosome. The ribosomes in prokaryotic cells have the same function as, but are smaller than, the ribosomes of eukaryotic cells.

Now that we know the ultrastructure of prokaryotic and eukaryotic cells, we can revisit Table 1.1. At the start of the chapter, we had not learned about cell ultrastructure. We can use our knowledge of cell ultrastructure to improve the comparison of prokaryotic cells and eukaryotic cells given in Table 1.1. This has been done in Table 1.5.

You should remember the information in Table 1.5 if asked to compare prokaryotic and eukaryotic cells in a Unit Test.

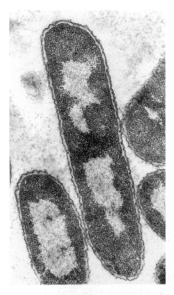

Figure 1.7
Electron micrograph of a bacterial cell. Notice that the genetic material is free in the cytoplasm and there are no cell organelles.

Feature	Prokaryotic cell	Eukaryotic cell
Capsule	Present	Absent
Cell wall	Present (murein)	Present in fungi (chitin) and in plants (cellulose) Absent in animals
Plasma membrane	Present	Present
Cytoplasm	Present	Present
• ribosomes	• small ribosomes are present – they are always free in the cytoplasm	• larger ribosomes are present – some are free in the cytoplasm and some are attached to rough ER
• other organelles	• absent	• present – they include chloroplasts, ER, Golgi apparatus, lysosomes and mitochondria
Nucleus	Absent	Present
• nuclear envelope	• absent	• present
• nucleoli	• absent	• present
• chromosomes	• single and circular	• many and linear

Table 1.5 A fuller comparison of prokaryotic and eukaryotic cells than that given in Table 1.1.

Cell fractionation and ultracentrifugation: separating cell organelles

You might wonder how biologists have been able to work out the function of organelles, given their size. This is done by collecting a sample containing large numbers of only one type of organelle. The function of the organelles in such a sample can then be studied. Samples of only one organelle can be obtained by the process of cell fractionation followed by the process of cell ultracentrifugation.

Cell fractionation breaks up (homogenises) cells. This can be done using a pestle and mortar. (You might use a pestle and mortar at home, for example to grind spices together.) Electric blenders and homogenisers are more likely to be used in research laboratories for breaking up cells. After breaking up the cells, a fluid mixture called a **homogenate** is obtained. This homogenate is filtered to remove bits of cells that have not been broken up properly. The filtrate from this process contains a mixture of cell organelles. We now need to separate the different organelles in this mixture.

Q 6 Whilst being homogenised, cells are kept in a solution that:
(a) has the same water potential as the cells
(b) has a constant pH
(c) is cold.
Suggest a reason for each of (a), (b) and (c).

Ultracentrifugation separates the components of cells that have been broken up. The separation is done using a centrifuge. This is an instrument that spins tubes of fluid around at high speeds. As they spin, solid particles in the fluid are thrown to the bottom of the tube, forming a pellet. We need to produce pellets with only one type of organelle. We can ensure which type of organelle is found in the pellet by controlling:

● the speed at which the centrifuge spins

● the time for which the centrifuge spins.

The flowchart in Figure 1.8 summarises the stages of cell fractionation and ultracentrifugation.

Figure 1.8
The stages of cell fractionation and ultracentrifugation.

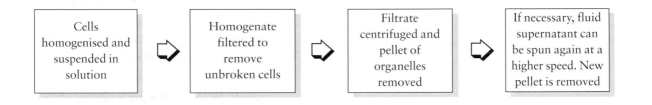

Table 1.6 shows the times and speeds used to produce pure pellets containing different types of organelle. As you might expect, the larger components form pellets at slower speeds than smaller components.

Speed of centrifugation (g)	Time of centrifugation (minutes)	Organelles in pellet
500 to 1000	10	Nuclei and chloroplasts
10 000 to 20 000	20	Mitochondria and lysosomes
100 000	60	Rough ER and ribosomes

Table 1.6 Ultracentrifugation of cell homogenate to obtain samples of organelles (g is the acceleration due to gravity, which has a value of $9.8 \ m \ s^{-2}$).

Q 7 Which organelles would remain in the supernatant if an animal cell homogenate was centrifuged at 800g for 10 minutes? How would you obtain a pure sample of the smallest of these organelles?

Summary

- Cells are the smallest unit of structure in organisms.

- Cells can be seen using microscopes. A microscope enlarges the image of cells. This is called magnification. To show us small structures within cells, a microscope must also be able to distinguish between objects that are close together. This is called resolution.

- An optical microscope passes light through a specimen to be viewed. Its maximum magnification is about 1500 times and its limit of resolution is about 2 μm.

- An electron microscope uses beams of electrons. Its maximum magnification is about 500 000 times and its limit of resolution is about 1 nm. A transmission electron microscope passes beams of electrons through a specimen to be viewed. A scanning electron microscope bounces beams of electrons off the surface of the specimen to be viewed.

- Prokaryotic cells do not have a true nucleus. Only bacteria have prokaryotic cells.

- Eukaryotic cells have a nucleus that is separated from the cytoplasm by a nuclear envelope. Organisms other than bacteria have eukaryotic cells.

- Prokaryotic cells are surrounded by a cell wall and a plasma membrane. Some are surrounded by a capsule. Ribosomes are the only organelles found in their cytoplasm.

- Eukaryotic cells are surrounded by a plasma membrane. Some are surrounded by a cell wall (e.g. fungi and plants). Their cytoplasm contains a variety of organelles.

- Each organelle has a particular function. Mitochondria make ATP. Chloroplasts make carbohydrates in photosynthesising cells. Ribosomes make proteins. Endoplasmic reticulum forms channels through cytoplasm that help to distribute substances quickly around cells. Golgi apparatus, found in some cells, stores substances that cells have made before they are carried to the surface of the cell by small vesicles. Lysosomes are vesicles that contain protein-digesting enzymes.

- Cells have at least one chromosome that controls their activities. Prokaryotic cells have a single, circular chromosome that is in their cytoplasm. Eukaryotic cells have several linear chromosomes that are found in the nucleus.

- Cell organelles must be collected from cells before we can study their function. This is done by breaking cells into their components (cell fractionation) and then separating these components using a centrifuge (ultracentrifugation).

Assignment

Once you have read this chapter you will appreciate that although cells are very small, we have many tools and techniques available to examine them and find out more about their structure. But we have a problem. Almost everything we do to a cell alters it in some way. Because we must be sure that what we are looking at through a microscope is what we see in real life, biologists need to be able to interpret images correctly. In this assignment we shall look at one very specialised cell, a human red blood cell. We shall look at what we can find out using different types of microscope and show how we need to be careful in interpreting what we see.

In order to look at blood cells with an optical microscope, we can make a blood smear. The technique for doing this is shown in Figure 1.9. The photograph in Figure 1.10 shows the appearance of some of the red blood cells from this smear when seen with an optical microscope. Look at these cells carefully and then answer the following questions.

2 A second slide is allowed to touch the blood. This slide is then pulled along in the direction shown by the arrow. It leaves a thin smear of blood behind it.

Figure 1.9
Making a thin smear of blood for examining with an optical microscope.

1 A small drop of blood is placed at one end of a microscope slide.

1 The shape of a red blood cell is often described as a biconcave disc. What is the evidence from this photograph (Figure 1.10) that these cells could have a biconcave shape?

(1 mark)

2 The magnification of a cell is the size it appears in a photograph or drawing, divided by its actual size. This can be written as a simple formula:

$$\text{magnification} = \frac{\text{apparent size of cell}}{\text{actual size of cell}}$$

The magnification of the photograph in Figure 1.10 is $\times 1000$. Calculate the actual diameter of a human red blood cell in micrometres. Show your working.

(2 marks)

A red blood cell is surrounded by a plasma membrane. We can find out something about this membrane by treating the cells in different ways before examining them.

3 When the slide is waved backwards and forwards and dried rapidly before it is examined, the cells appear round and smooth, as in the photograph. When, however, the slide is left on the bench to dry slowly, the cells appear smaller and crinkled around their edges.

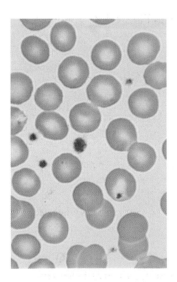

Figure 1.10
Red blood cells seen with an optical microscope.

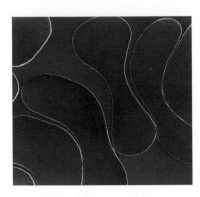

Figure 1.11
Red blood cells seen with a
transmission electron
microscope.

Figure 1.12
Sections have been cut through
this red blood cell in different
planes.

Suggest an explanation for the different appearance of the cells when they have dried out slowly.

(2 marks)

4 Detergents dissolve lipids. If a drop of detergent is added to the blood on the slide before it is examined, no cells can be seen at all. What does this suggest about the plasma membrane surrounding the red blood cell?

(1 mark)

Using an optical microscope of the sort found in a school or college laboratory, it is possible to find out a lot about red blood cells and the membranes that surround them. We can find out even more if we examine them with an electron microscope.

Look at Figure 1.11, which shows red blood cells as seen with a transmission electron microscope. In a transmission electron microscope, electrons pass through the specimen. It is not possible to look at a whole cell because it is much too thick to allow the electrons to pass through, so a thin section must be cut. Figure 1.12 shows a red blood cell that has been cut through in different planes.

5 (a) Make simple drawings to show the appearance of the cell if it were cut through each of the planes shown in Figure 1.12.

(3 marks)

(b) Suggest an explanation for the different shapes of the red blood cells shown in Figure 1.11.

(1 mark)

6 What evidence is there that the photograph in Figure 1.11 has been taken through an electron microscope?

(1 mark)

7 Rewrite the following table, matching each conclusion with the appropriate piece of evidence.

Conclusion	Evidence
The plasma membrane allows water molecules to pass through it	There are no mitochondria present
Many of the molecules in the cytoplasm are of the same type	The appearance of the cells when seen with a scanning electron microscope
Red blood cells do not use oxygen and cannot respire aerobically	The red blood cells in a smear that is left to dry slowly appear smaller and crinkled around the edges
Red blood cells are biconcave in shape	There is no nucleus present
Red blood cells do not contain DNA, so they cannot make proteins	The red blood cells in Figure 1.11 are a uniform dark colour

Table 1.7

(3 marks)

Examination questions

1 (a) Table 1.8 shows some features of cells. Complete the table with ticks to show those features that are present in an epithelial cell lining the small intestine and those which may be present in a prokaryotic cell

Feature	Epithelial cell from small intestine	Prokaryotic cell
Golgi apparatus		
Mitochondrion		
Nuclear envelope		
Plasmid		
Ribosome		

Table 1.8

(2 marks)

(b) (i) Explain why it is possible to see the detailed structure of a cell with an electron microscope but not with an optical microscope.

(2 marks)

(ii) Care must be taken when interpreting electronmicrographs. Some features seen using an electron microscope might not be present in the living cell. Suggest an explanation for this observation.

(1 mark)

Figure 1.13

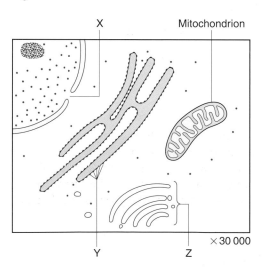

2 Figure 1.13 shows part of an animal cell.

(a) Name feature X.

(1 mark)

(b) Describe the function of organelle Y.

(1 mark)

(c) Describe one way in which the function of organelle Z is related to the function of organelle Y.

(1 mark)

Calculate the actual length of the mitochondrion in millimetres. Show your working.

(2 marks)

Getting In and Out of Cells

Human cells need a constant supply of glucose to remain healthy. People who suffer from diabetes mellitus are unable to keep a constant supply of glucose in their cells. These people are called diabetics.

The British Diabetics Association estimates that 1.4 million people in the UK have been diagnosed with diabetes mellitus. The Association also estimates that a further 1 million people in the UK have diabetes mellitus but do not yet realise it.

Keeping a constant supply of glucose within cells is a complex process. Part of this process depends on a hormone called insulin. Many people are diabetic because they cannot produce insulin: they suffer from Type 1 diabetes. The boy in Figure 2.1 suffers from Type 1 diabetes. He has to inject himself regularly with insulin because he cannot produce it. Over 75% of diabetics, however, can produce insulin: they suffer from Type 2 diabetes.

Type 2 diabetes is caused by faults in plasma membranes. The plasma membranes of some sufferers can no longer 'recognise' insulin. The plasma membranes of other sufferers can 'recognise' insulin but have lost the ability to increase the movement of glucose through themselves.

The types of diabetes mellitus illustrate three ways in which plasma membranes act. In the rest of this chapter you will learn how cells use their plasma membranes to:

- 'recognise' chemical messengers, such as insulin

- control the movement of substances into the cell, such as glucose

- release substances that cells have made, such as insulin.

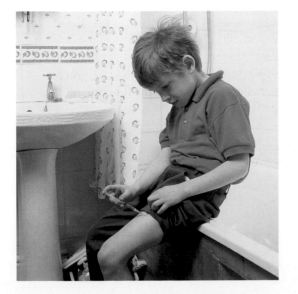

Cells and their surroundings

All cells are surrounded by a plasma membrane. A cell must get the raw materials it needs from its surroundings through its plasma membrane. A cell must also excrete the waste products it has made into the cell's surroundings through the plasma membrane.

Figure 2.1
This boy cannot produce a hormone called insulin. If he is to stay alive, he must inject himself with insulin each day.

Q 1 Name one raw material that every cell in your body must have and one waste product that they all must excrete.

Substances get through membranes in a number of ways. These are:

- diffusion
- facilitated diffusion
- osmosis
- active transport
- endocytosis
- exocytosis.

Before we can understand these processes, we must know something about the structure of the membrane itself.

Structure of plasma membrane

Membrane occurs inside cells as well as around them. Look at Figure 2.2, which shows a simplified plant cell. Its membranes are shown in blue. Oxygen molecules are produced inside chloroplasts, e.g. at the point labelled A in Figure 2.2. Work out from the diagram how many layers of membrane a molecule of oxygen must pass through to get out of the cell from the point labelled A.

Figure 2.2
A simplified plant cell with the membrane shown in blue. Through how many layers of membrane must a molecule of oxygen produced at point A pass in order to get out of the cell?

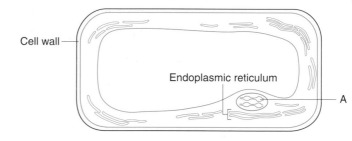

All the blue lines in Figure 2.2 represent membrane. Whether membrane is at the surface of the cell or in its cytoplasm, it has the same structure. Figure 2.3 shows what we think this structure is. It is described as the **fluid-mosaic model** of membrane structure.

Figure 2.3
The fluid-mosaic model of membrane structure.

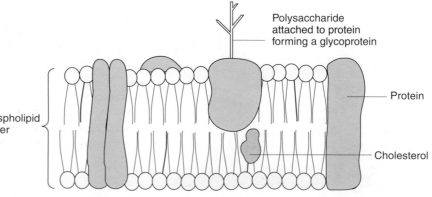

key term

Fluid-mosaic model
A membrane is a phospholipid bilayer studded with proteins, polysaccharides and other sorts of lipids. This patchwork of molecules is like a mosaic. Because the molecules move about within their respective layer, the membrane behaves like a fluid. This is why the model is called the fluid-mosaic model.

Q 2 Use the information in Figure 2.2 to suggest why we cannot see the structure of a membrane and need to produce a model of its probable structure.

A membrane is a **phospholipid bilayer**. This means that it has two layers of molecules called phospholipids. Each of these phospholipid molecules has two parts:

- a 'head' that will mix with water but not with fat (i.e. it is **hydrophilic**)

- two 'tails' that will mix with fat but not with water (i.e. they are **hydrophobic**).

In the phospholipid bilayer, the hydrophilic heads are always on the outside of the membrane. The hydrophobic tails are always on the inside of the membrane. Alone, this arrangement of phospholipids would form a barrier to water and to water-soluble substances. However, other molecules are scattered among the phospholipids. These include lipids (including cholesterol in the membranes of animals), proteins and polysaccharides. You will learn more about the structure of these substances in Chapter 3.

Q 3 Suggest why the hydrophilic heads are always on the outside of a phospholipid bilayer and the hydrophobic tails are always on the inside.

The proteins in membranes are of special interest to us. This is because they have a number of important functions. Proteins function as:

- **carriers** for water-soluble molecules (such as glucose)

- **channels** for ions (such as sodium and chloride ions)

- **pumps,** which use energy to move water-soluble molecules and ions

- **receptors,** which enable hormones and nerve transmitters to bind to specific cells

- **recognition sites,** which identify a cell as being of a particular type

- **enzymes,** which speed up chemical reactions at the edge of the membrane

- **adhesion sites,** which help some cells to stick together.

Polysaccharides are bound to some of the lipids and proteins studded in the phospholipid bilayer. Proteins that are bound to polysaccharides are called glycoproteins. It is these glycoproteins in the membrane that act as receptors and recognition sites.

Q 4 Cholesterol, found in animal cell membranes, reduces the sideways movement of phospholipid molecules within a membrane. Suggest why the presence of cholesterol might reduce the rate of movement of molecules through a membrane.

Diffusion

Because molecules and ions have inherent energy, they constantly move about. Figure 2.4 shows the effects of the movement of ink molecules and water molecules. In beaker A you can see a droplet of ink that has just been placed in water. Beaker A illustrates an important concept: that of the **concentration gradient**. 'Concentration' refers to the number of molecules or ions in a specified volume. 'Gradient' shows that one region contains more molecules or ions than a neighbouring region. In beaker A, there are more molecules of ink in the droplet than in the surrounding water. As a result, a concentration gradient occurs between the region containing the droplet and the surrounding region containing only water.

Figure 2.4
The effect of diffusion. A drop of ink was placed in the left-hand beaker of water. After a few hours, movement of ink molecules and water molecules caused the appearance of the right-hand beaker. By now, the molecules of ink and water are evenly spaced throughout the ink solution.

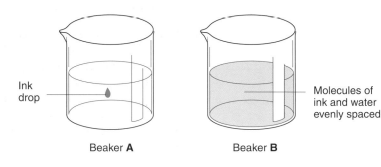

Ink drop

Molecules of ink and water evenly spaced

Beaker **A**   Beaker **B**

key term

Concentration gradient

A concentration gradient occurs when one region contains more of a molecule or ion than a neighbouring region. Molecules and ions move down a concentration gradient. When the concentration difference is large, the net movement of molecules or ions down the gradient is faster than when the concentration difference is small.

Beaker B in Figure 2.4 shows what happened when beaker A was left untouched. After a few hours, movement of the ink molecules has spread them evenly throughout the water of the beaker. During this time, ink molecules moved in all directions and they continue to do so. However, because they are now evenly spread out, more must have moved from where they were concentrated in the droplet to where they were less concentrated in the water. In other words, their net movement has been down the concentration gradient. This is the process of diffusion.

key term

Diffusion

Diffusion is the net movement of molecules or ions from a region of higher concentration to a region of lower concentration. (This can also be described as the net movement of molecules or ions down a concentration gradient.)

Extension box 1

More about gradients

In the ink solution described in the text, the dissolved ink is called the **solute** and the water in which the ink is dissolved is called the **solvent**. The cytoplasm of cells is mainly water, so water is always the solvent in cell solutions. However, there are many different solutes in cells.

If more than one solute is present in a solution, each solute still diffuses along its own concentration gradient. For example, in a human liver cell it is likely that glucose molecules will diffuse into the cell along a glucose concentration gradient. At the same time it is likely that urea molecules will diffuse out of the cell along a urea concentration gradient. The two concentration gradients do not interfere with each other.

The process of diffusion described so far applies to uncharged molecules. When a solute is charged, as ions are, diffusion occurs along an **electric gradient**. An electric gradient is caused by two adjacent regions having different electrical charges. If the electric gradient is strong enough, diffusion of ions can occur even against their own concentration gradient. Processes such as the flow of impulses in your nervous system depend on the balance of diffusion along both electric and concentration gradients.

Differences in pressure result in pressure gradients. These gradients can also influence the speed and direction in which solute molecules diffuse. You will see in Chapter 6 how the movement of molecules and ions between blood capillaries and the cells they supply is affected by pressure gradients.

Figure 2.5
Cells of the same type form tissues. Diffusion across the thin cells in tissue A will be faster than across the thicker cells in tissue B. Although they are the same shape as cells in tissue B, the extension of the membrane of cells in tissue C increases their surface area. What effect will this have on the rate of diffusion across C compared with B?

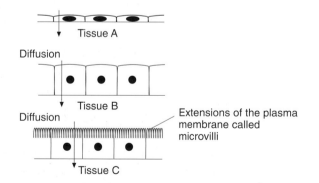

The rate of diffusion is affected by a number of factors. One of these, differences in concentration, has already been described above. In biology, we are usually interested in the rate of diffusion across plasma membranes or across epithelia (see Chapter 5). These are often called **exchange surfaces**. Two further factors influence the rate of diffusion across exchange surfaces:

- the surface area across which diffusion occurs

- the thickness of the surface.

Figure 2.5 shows three groups of cells that line various parts of your body. The cells in A are thinner than those in B. As a result, diffusion will normally be faster across A than across B. The cells in C are similar in shape to those in B. However, the cells in C have microscopic extensions of their plasma membrane. As a result of this increase in surface area, diffusion will normally be faster across C than across B.

Q 5 Under what conditions might the diffusion rate across B in Figure 2.5 be faster than across A? Explain your answer.

We now have three factors affecting the rate of diffusion. Diffusion through an exchange surface is:

● faster when the difference in concentration is high (i.e. the concentration gradient is steep)

● faster when the exchange surface has a large area

● slower when the exchange surface is thick.

The relationship between these three factors is described by Fick's law. Although this law is shown as an equation, you will not be expected to use it to perform calculations.

key term

Fick's law   The rate of diffusion through an exchange surface is proportional to:

$$\frac{\text{surface area} \times \text{difference in concentration}}{\text{thickness of surface}}$$

Q 6 Use Fick's law to explain the rates of diffusion across the exchange surfaces in Figure 2.5.

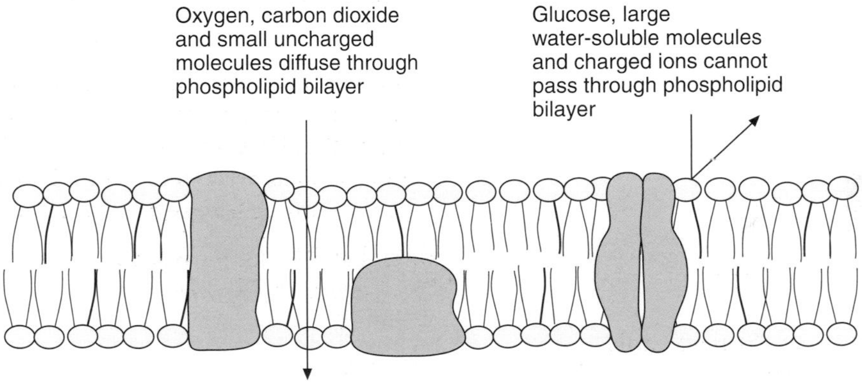

Oxygen, carbon dioxide and small uncharged molecules diffuse through phospholipid bilayer

Glucose, large water-soluble molecules and charged ions cannot pass through phospholipid bilayer

Figure 2.6
Oxygen, carbon dioxide and other small uncharged molecules can diffuse freely across the phospholipid bilayer of a plasma membrane. Glucose, other large water-soluble molecules and charged ions cannot.

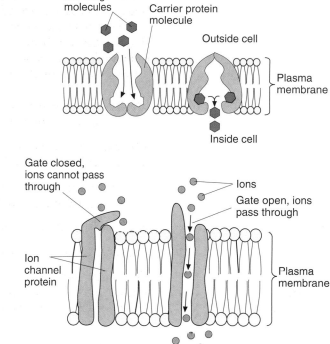

Figure 2.7
The two types of protein involved in facilitated diffusion. Carrier proteins bind to the diffusing molecule and take it through the membrane by changing shape. Ion channels help the diffusion of charged ions. Note that some ion channels have gates that open and close.

Q 7 Suggest why
(a) **lipid-soluble molecules** and
(b) **small molecules can diffuse through plasma membranes.**

Facilitated diffusion

Oxygen and carbon dioxide are small, uncharged molecules. They diffuse freely across the phospholipid bilayer of a cell's plasma membrane (Figure 2.6). We could say that the membrane is no barrier to their diffusion.

In contrast, large molecules, such as glucose, almost never diffuse across the bilayer. Neither do charged ions. Instead, these substances must cross the membrane through proteins that span both sides of the bilayer. Figure 2.7 shows the two types of protein that are involved in facilitated diffusion:

- **carrier proteins** bind to a specific type of diffusing molecule. When they have done so, they change shape, releasing the diffusing molecule at the other side of the membrane

- **ion channels** are formed by proteins with a central pore that is lined with charged groups. The diffusion of charged particles, such as Ca^{2+}, Na^+, K^+, HCO_3^- and Cl^- ions, is helped by these ion channels. Some ion channels are gated (Figure 2.7) and so can open or close. These gates allow cells to regulate the flow of ions from one cell to another.

Note that, like simple diffusion, neither of these processes involves the use of energy by the cell.

Osmosis

We saw earlier in this chapter that a plasma membrane lets molecules and ions pass through itself. However, many types of large molecule cannot diffuse through a plasma membrane. Because the membrane only allows the passage of some molecules, it is called **partially permeable.**

key term

Partially permeable membrane

A partially permeable membrane will only let small molecules pass through it.

The cytoplasm of all cells contains large numbers of protein molecules that cannot diffuse out of the cells. These molecules become concentrated in the cytoplasm of the cell. As a result, a protein concentration gradient exists across cell surface membranes. This gradient affects the diffusion of water molecules across membranes. Figure 2.8 shows how this happens.

Figure 2.8
A model to show osmosis. Solution A is a weak solution of protein. Its water potential has a negative value, −3 kPa. This solution has a low concentration of protein molecules but a high concentration of water molecules. Solution B is a more concentrated protein solution. Its water potential has a more negative value than solution A, −7 kPa. Solution B has a higher concentration of protein molecules but a lower concentration of water molecules than solution A. Solution A is separated from solution B by a partially permeable membrane. As a result of osmosis, there will be a net movement of water from solution A to solution B through the partially permeable membrane.

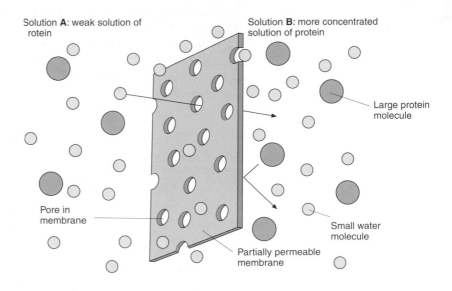

On the left of Figure 2.8 there is a weak protein solution. On the right of Figure 2.8 is a stronger protein solution. The two are separated by a partially permeable membrane. Because of their inherent energy, water molecules will constantly move about. However, there will be a net movement of water molecules from the weaker solution to the stronger solution. In other words, water molecules will diffuse along the water concentration gradient. The protein molecules will also move about. They cannot diffuse along their concentration gradient because they cannot pass through the partially permeable membrane. This movement of water molecules happens with other solutes as well as protein and is called osmosis. In osmosis, there is a net movement of water molecules through a partially permeable membrane from where they are concentrated (i.e. a weak solution) to where they are less concentrated (i.e. a concentrated solution).

key term

Osmosis

Osmosis relates to the diffusion of water molecules. It is the net movement of water through a partially permeable membrane from a solution of less negative water potential to a solution of more negative water potential. (The water potential of pure water is zero. All solutions have a water potential with a negative value.)

Water potential

We can describe the movement of water molecules in Figure 2.8 in another way. This involves the use of the term **water potential**. Water potential is the ability of a solution to absorb water molecules by osmosis. It is measured in units of pressure called kilopascals (kPa). Pure water cannot absorb any more water by osmosis: it has a water potential of zero (0 kPa). A solution can absorb more water by osmosis. A solution always has a negative water potential (e.g. −6 kPa). The stronger the solution, the more negative its water potential. Using the concept of water potential, we can redefine osmosis as the net movement of water through a partially permeable membrane from a solution of less negative water potential to a solution of more negative water potential.

Q 8 In each case, state the direction in which water will move by osmosis.
   (a) solution P with a water potential of −1 kPa and solution Q with a water potential of −4 kPa
   (b) solution R with a water potential of −1 kPa and solution S with a water potential of 0 kPa.

Osmosis in cells

Osmosis has a significant effect on cells. Figure 2.9 shows what happens to human red blood cells when they are placed in solutions with different water potentials. In a solution with a more negative water potential than themselves, they lose water by osmosis and shrink. In a solution with a less negative water potential than themselves, they gain water by osmosis and eventually burst. Figure 2.10 shows that a similar process happens in plant cells when they are placed in solutions with different water potentials. Unlike red blood cells, plant cells have a cellulose wall. Because this wall is strong, it pushes on expanding cells. This stops them bursting in a solution of less negative water potential than their own cytoplasm.

Red blood cells in solution
with same water potential
as the cell cytoplasm

Cell in solution
with more negative
water potential

Cell in solution
with less negative
water potential

Cell smaller and
appears 'crinkled'

Cell swells and
bursts

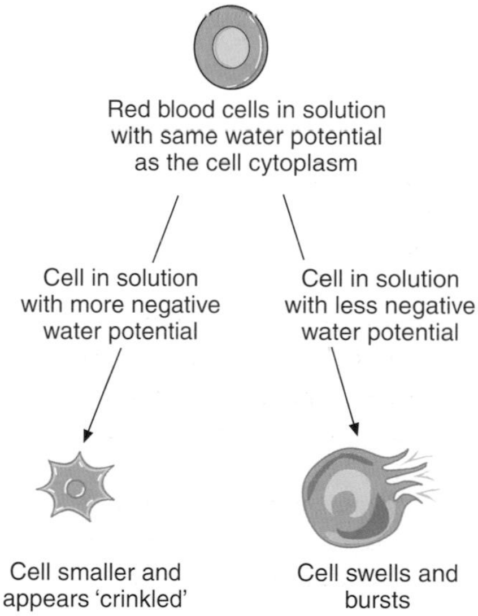

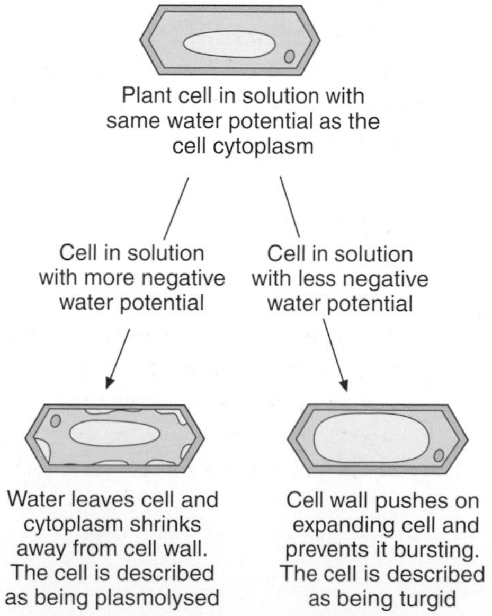

Figure 2.9 (above)
The effects of osmosis on human red blood cells. All animal cells are affected in this way.

Figure 2.10 (right)
The effects of osmosis on plant cells. Because the wall is able to press onto the expanding cell it stops these cells bursting in solutions with a less negative water potential than their own cytoplasm.

Plant cell in solution with
same water potential as the
cell cytoplasm

Cell in solution
with more negative
water potential

Cell in solution
with less negative
water potential

Water leaves cell and
cytoplasm shrinks
away from cell wall.
The cell is described
as being plasmolysed

Cell wall pushes on
expanding cell and
prevents it bursting.
The cell is described
as being turgid

Figure 2.11
Active transport involves carrier proteins in the phospholipid bilayer of a membrane. The carrier molecules can only change shape following an input of energy from the cell.

Active transport

Diffusion, facilitated diffusion and osmosis depend on the energy inherent to molecules and ions. Cells do not need to use energy when these processes occur. Because of this, these processes are described as **passive transport**. Sometimes cells move substances against their concentration gradient. In this case, cells need to use energy to make substances cross their membranes. Because of this, this process is called **active transport**.

Figure 2.11 shows that active transport involves the use of carrier proteins in the phospholipid bilayer of membrane. The substance to be transported binds to a carrier protein. This protein changes shape, releasing the substance to be transported on the opposite side of the membrane. So far, this is similar to facilitated diffusion. However, in active transport the carrier protein needs to be activated by an input of energy from the cell.

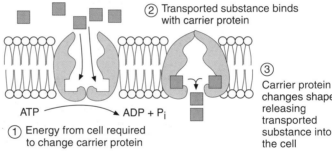

② Transported substance binds with carrier protein

③ Carrier protein changes shape, releasing transported substance into the cell

ATP ⟶ ADP + P$_i$

① Energy from cell required to change carrier protein

Extension box 2

Energy for active transport

The energy used in active transport is released when a molecule of adenosine triphosphate (ATP) is broken down in cells. You will learn more about this reaction if you study Module 5 of the AQA specification. For the time being, all we need to know is that adenosine triphosphate is broken down to form adenosine diphosphate (ADP) and an inorganic phosphate group (Pi). The reaction involves an enzyme called ATPase. We can represent the reaction in a simple equation.

$$\text{ATP} \xrightarrow{\text{ATPase}} \text{ADP} + \text{Pi} + \text{energy}$$

Cells continuously produce ATP in a process called respiration. For this reason, the rate of active transport is affected by the rate of cell respiration. It is also slowed by poisons such as cyanide that slow respiration.

Endocytosis and exocytosis

Passive and active transport move only one to a few ions or molecules at a time. Endocytosis and exocytosis move substances across membranes in bulk. To do this, they use small membrane-bound sacs called vesicles. These vesicles are formed when they 'bud off' another membrane. Figure 2.12 shows both endocytosis and exocytosis.

Figure 2.12
An overview of (a) endocytosis and (b) exocytosis.

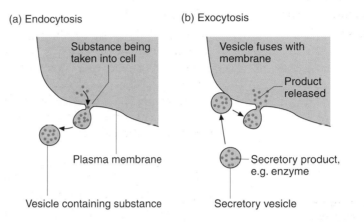

In **endocytosis,** part of the plasma membrane sinks into the cell. It then 'buds off' and seals back onto itself. This produces a vesicle that contains substances from outside the cell. Endocytosis that brings solid material into the cell is called **phagocytosis.** Endocytosis that brings fluid into the cell is called **pinocytosis.** In **exocytosis,** a vesicle is formed in the cytoplasm. For example, a vesicle might 'bud off' from the edge of Golgi apparatus in the cytoplasm. Once formed, the vesicle moves to the plasma membrane. It then fuses with the plasma membrane so that its contents are pushed outside the cell. Hormones such as insulin (see start of chapter) are secreted from cells in this way.

Summary

- The membrane surrounding a cell is called the plasma membrane.

- The fluid-mosaic model of membrane structure shows that membranes have a phospholipid bilayer. This bilayer is studded with proteins and polysaccharides.

- Proteins in the plasma membrane perform a variety of functions. They act as cell receptors, as recognition sites and as adhesion sites. Proteins also assist the movement of molecules or ions through membranes.

- Diffusion is the net movement of molecules or ions down a concentration gradient. According to Fick's law, diffusion across exchange surfaces is fastest when the exchange surface has a large area, the exchange surface is thin and the concentration gradient is steep.

- Diffusion is sometimes helped by specific proteins that form part of the structure of membrane. This process is called facilitated diffusion.

- Water potential is the ability of a solution to absorb water molecules by osmosis. It is measured in units of pressure called kilopascals (kPa). Pure water has a water potential of zero (0 kPa). A solution always has a negative water potential (e.g. −6 kPa).

- Osmosis is the diffusion of water. It is the net movement of water through a partially permeable membrane from a solution of less negative water potential to a solution of more negative water potential.

- Diffusion, facilitated diffusion and osmosis depend on the energy inherent in molecules and ions. Because the cell does not use energy in these processes, they are referred to as passive transport.

- Active transport is a process by which cells move molecules or ions against their concentration gradient. It involves the use of energy by cells. Like facilitated diffusion, this process also involves carrier proteins in membranes.

- Endocytosis is the bulk movement of substances into a cell. In this case, substances become trapped in a vesicle that 'buds off' from the plasma membrane. In phagocytosis the trapped substances are in solid form and in pinocytosis the trapped substances are in fluid form.

- Exocytosis is the bulk movement of substances out of a cell. These substances become trapped inside a vesicle formed in the cytoplasm. They are secreted when the vesicle moves to the plasma membrane and fuses with it.

Assignment

One of the skills that you need as a biologist is the ability to apply your knowledge to new situations. You should be able to use your understanding of basic principles to explain something that, at first sight, seems unfamiliar.

This assignment consists of a passage about the treatment of kidney disease. Read it carefully then answer the questions that follow. You will need to use information in this chapter about the way in which different substances get into and out of cells. You will also require some of your knowledge of cell structure (Chapter 1).

Figure 2.13
This patient has severe kidney disease and her blood is being treated by passing it through an artificial kidney.

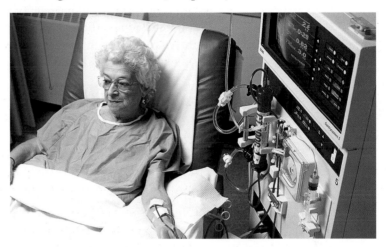

An adult human kidney weighs approximately 100 g and is about the size of a clenched fist. It contains a million or so tiny tubes called kidney tubules or nephrons. In the first part of one of these nephrons, much of the water and soluble substances contained in the blood are
5   filtered out and pass into the cavity of the nephron. As this filtrate passes down the tubule, all of the glucose, which is useful to the body, is reabsorbed and goes back into the blood. Waste products such as urea remain in the filtrate. A lot of the water present in the filtrate is also reabsorbed, leaving a concentrated solution of waste substances,
10 urine.

Kidneys are vital. If they stop working, and the person is left untreated, a build up of waste products in the body rapidly results in death. However, patients whose kidneys no longer function effectively can be treated with a kidney machine or artificial kidney.
15 Two or three times a week, a fine needle is inserted into a blood vessel in the patient's arm or leg and blood is pumped through the artificial kidney and back to the patient. The most important part of this artificial kidney is the dialyser. This contains a large number of small diameter tubes made from partially permeable membrane. Dialysis
20 fluid is pumped through these tubes. The fluid contains specific concentrations of sodium, potassium and chloride ions and glucose, but it does not contain any urea. Waste products and excess ions pass

from the blood through the membrane and into the dialysis fluid. Waste dialysis fluid is removed.

25 There is another, more recently developed, method of dialysis. This uses the patient's own body as an artificial kidney. A small cut is made in the body wall just above the navel and a tube is inserted into the body cavity. Dialysis fluid is poured into the abdomen through this tube. Once this has been done, a cap is placed over the end of the tube

30 and the patient can continue with whatever he or she was doing and move around normally. The dialysis fluid in the cavity of the abdomen is separated from the blood supply by a partially permeable membrane called the peritoneum. This surrounds the organs in the abdominal cavity. Waste products pass from the blood, through the

35 peritoneum into the dialysis fluid. Excess water is removed from the blood by putting additional glucose in the dialysis fluid. After several hours, the waste dialysis fluid is drained out of the abdominal cavity into a plastic bag and disposed of.

Now answer the questions below. The first two relate to the way in which a healthy kidney works (described in the first paragraph, lines 3–10). Marks have been added after each question. These should help you to decide on the amount of detail you are required to give in your answer.

1 Figure 2.14 shows a kidney tubule.

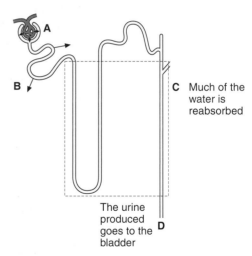

Figure 2.14
A nephron and the way in which it works.

Use information in the passage to write suitable labels explaining what happens at points A and B.

(2 marks)

2 (a) Under normal body conditions, all the glucose in the filtrate in the kidney is reabsorbed back into the blood (lines 6–7). Giving a reason for your answer, explain whether you think active transport, diffusion or osmosis are involved in the reabsorption of this glucose.

(b) Glucose is reabsorbed in the first part of the nephron. Figure 2.15 shows one of the cells that form the wall of this part of a nephron.

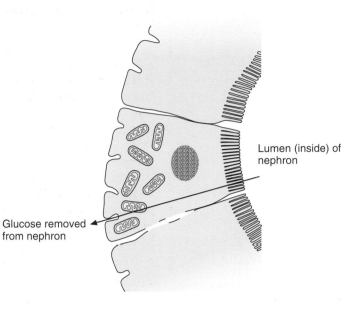

Figure 2.15
A cell from the wall of the first part of a nephron.

Lumen (inside) of nephron

Glucose removed from nephron

Explain how two of the features that can be seen in this cell are involved in the process by which glucose is reabsorbed.

(3 marks)

The artificial kidney described in Paragraph 2 depends on diffusion to remove waste products and excess ions from the blood. Your knowledge of this topic will help you to understand the way in which an artificial kidney works.

3 (a) Explain how diffusion enables the waste product, urea, to be removed from the blood in an artificial kidney.

(2 marks)

(b) Excess ions are removed but some remain in the blood. Explain why the dialyser only removes excess ions.

(2 marks)

4 Explain how the design of the dialyser prevents red blood cells and large molecules such as proteins being removed from the blood.

(2 marks)

5 Suggest the advantage of having a dialyser that contains a large number of small tubes rather than a smaller number of much larger tubes.

(2 marks)

Questions 6 and 7 relate to the method of dialysis described in Paragraph 3.

6 This method of dialysis is called continuous ambulatory peritoneal dialysis (CAPD). Explain why it is given this name.

(3 marks)

7 The passage refers to putting additional glucose in the dialysis fluid in order to remove excess water from the blood (lines 35–36). Explain why adding extra glucose will allow water to be removed from the blood.

(2 marks)

Examination questions

Figure 2.16

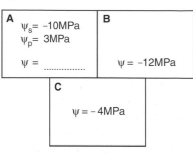

Figure 2.17

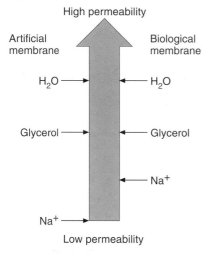

Figure 2.18

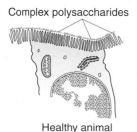

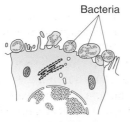

1 (a) Figure 2.16 shows three adjacent plant cells.

  (i) The water potential of a plant cell (ψ) is affected by the water potential of the solutes in the cytoplasm (ψ_s) and the pressure caused by the cell wall (ψ_p). Use the formula

  $$\psi = \psi_s + \psi_p$$

  to calculate the water potential of cell A.

  (1 mark)

  (ii) Put arrows on the diagrams to show the movement of water between these cells.

  (1 mark)

  (iii) Explain why the water potential of a sucrose solution has a negative value.

  (2 marks)

  (b) It is possible to make an artificial membrane which has only a lipid bilayer. Figure 2.17 compares the permeability of such an artificial membrane with a biological cell membrane.

  (i) Explain why the permeability to glycerol is the same in both membranes.

  (1 mark)

  (ii) Explain why the permeability to sodium ions is different in the two membranes.

  (2 marks)

2 Figure 2.18 shows the surface of epithelial cells lining the small intestine. The first is from a healthy mammal. The second is from a mammal whose intestine has been invaded by disease-causing *Escherichia coli* bacteria.

  (a) (i) Give one piece of evidence from the drawings that shows that these cells were viewed with an electron microscope.

  (1 mark)

  (ii) Describe the effect of the *Escherichia coli* bacteria on the surface of the epithelial cells.

  (1 mark)

  (b) Use the diagram to describe and explain the effect the *Escherichia coli* bacteria would have on absorption of glucose by these cells.

  Description   *(1 mark)*
  Explanation   *(2 marks)*

Biological Molecules

In 1986, a new disease began to affect cattle. Animals that were apparently healthy would occasionally stumble as they moved around. Gradually their movements would become more and more uncoordinated until they were unable to stand. This disease became known as bovine spongiform encephalopathy or BSE for short: bovine because it affects cattle, spongiform because the brains of animals that die of BSE have a sponge-like appearance when examined with a microscope and encephalopathy because it is a condition that affects the brain (Figure 3.1).

Figure 3.1
BSE affected large numbers of cattle in the UK. Fears that it could lead to humans being infected led to a world-wide ban on the sale of British beef.

The agent responsible for BSE is very different from the microorganisms that cause many other infectious diseases. It is not a virus or a bacterium, but a protein molecule called a prion. Prions are found in the brain cells of healthy animals but we do not know what they do. A cow that develops BSE becomes infected with a foreign prion, probably by eating food contaminated with material obtained from sheep infected with a similar disease known as scrapie. These foreign prions are very like the normal prions in the cow's brain cells but their molecules differ in shape. Instead of being coiled into a helix, they are folded into sheets. The abnormal foreign prions cause the cow prions to change shape into these folded sheets. The sequence of events becomes a chain reaction, converting more and more of the normal prions into abnormal ones. The abnormal prions build up into fibres, which upset the function of the brain cells and eventually lead to their death.

Protein molecules such as prions are only one of many sorts of molecule found in the cells of living organisms. Cells contain many small molecules such as water, which makes up approximately 80% of the mass of a typical cell. In addition, they contain inorganic ions, such as those of calcium and sodium, that are essential for the cell to function. However, they also have a number of larger molecules that are built up from smaller chemical building blocks. This chapter will concentrate on these larger molecules.

Large molecules and small molecules

Molecules such as those of the carbohydrates, proteins and lipids found in living organisms are called organic molecules because they contain carbon. Carbon atoms are unusual in that they can form chemical bonds with other carbon atoms as well as with the atoms of different elements. Because of this many of the organic molecules found in living organisms are very large in size and are known as macromolecules. Many of these large molecules are made up of smaller molecules, which act as building blocks. Smaller building blocks that are identical or very similar to each other are known as monomers and they join together to form a polymer.

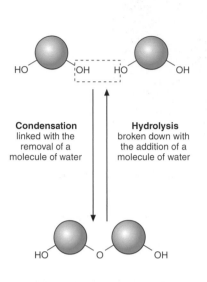

Condensation
linked with the removal of a molecule of water

Hydrolysis
broken down with the addition of a molecule of water

Figure 3.2
Monomers join together by condensation to form a polymer. This diagram shows two monomers joining together. When a large number of monomers are joined like this we get a polymer. A polymer can be broken down to its monomers by hydrolysis.

Q 1 **Starch, protein and fats are all large molecules. Starch is made up from many glucose molecules and proteins from a large number of amino acids. A fat consists of a molecule of glycerol and three fatty acid molecules. Which of the molecules mentioned is:**

(a) **a monomer**
(b) **a polymer**
(c) **a macromolecule?**

Figure 3.2 shows how two monomers can be joined together. This involves a chemical reaction known as condensation in which a molecule of water is formed. This water molecule comes from the hydrogen atom that is removed from one of these monomers and a hydroxyl (OH) group from the other. Because parts of the monomers have been removed, we refer to the parts that remain as residues once they have been joined. Joining a lot of monomers together in this way produces a polymer. Polymers may be broken down to the monomers from which they are formed by hydrolysis. This is the opposite reaction to condensation and involves the addition of water molecules.

Carbohydrates

A carbohydrate molecule contains carbon, hydrogen and oxygen. There are twice as many hydrogen atoms as oxygen atoms, the same proportion as in water. Carbohydrates are divided into three main types:

- monosaccharides are monomers. A monosaccharide is therefore a single sugar. Different monosaccharides contain different numbers of carbon atoms. Biologically important monosaccharides generally contain three (trioses), five (pentoses) or six (hexoses) carbon atoms

- disaccharides contain two sugar residues

- polysaccharides are very large molecules containing many sugar residues.

key term

Carbohydrates have the general formula $C_x(H_2O)_y$.

Figure 3.3
Carbohydrates play an important part in the biology of these orchids. Cellulose in the cell walls of the stems helps to provide support. As a result of photosynthesis, glucose is produced in the leaves. This is transported through the stem mainly as sucrose and stored as starch. The flowers produce a large amount of nectar, which is also rich in sucrose.

Table 3.1 summarises some information about these carbohydrates.

Type of carbohydrate	Examples	Made up from	Biological importance
Monosaccharide			
Triose	Triose		Intermediate product in the biochemical pathways of respiration and photosynthesis
Pentose	Ribose Deoxyribose		These sugars are found in the nucleic acids RNA and DNA
Hexose	Glucose Fructose		Important source of energy in respiration Found in many sweet-tasting fruits
Disaccharide	Sucrose	Glucose + fructose	The form in which carbohydrates are transported in plants
	Maltose	Glucose + glucose	Formed from the digestion of starch
	Lactose	Glucose + galactose	The carbohydrate found in milk
Polysaccharide	Starch	Glucose	The main storage carbohydrate in plants
	Glycogen	Glucose	The main storage carbohydrate in humans and other animals
	Cellulose	Glucose	An important component of plant cell walls

Table 3.1

Q 2 How many carbon atoms are there in a maltose molecule?

(a)

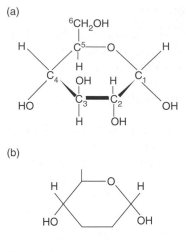

(b)

Figure 3.4
Diagram (a) shows the structural formula of a glucose molecule. The small numbers in red allow us to refer to particular carbon atoms. Diagram (b) is a simplified version of this and is all you need to learn.

Figure 3.5
The structural formulae of some hexose sugars. If you compare these with Figure 3.4, you will see that the molecule in Figure 3.4 is α-glucose.

Figure 3.6
Two α-glucose molecules may join together by condensation to give a molecule of the disaccharide maltose.

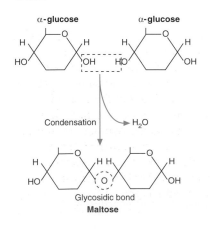

Glucose and other sugars

Glucose is a monosaccharide. It is also a hexose so a molecule of glucose contains six carbon atoms and has the molecular formula $C_6H_{12}O_6$. This formula simply tells us how many atoms are contained in the molecule; the structural formula shown in Figure 3.4 is more useful because it shows us how the atoms are arranged.

If you look at Table 3.1 again you will see that there are several different kinds of hexose sugar. All of them have the molecular formula $C_6H_{12}O_6$ but they differ from each other because their atoms are arranged in slightly different ways. Figure 3.5 shows the simplified structural formulae of four different hexose sugars. The slight differences in the way their atoms are arranged give them slightly different properties.

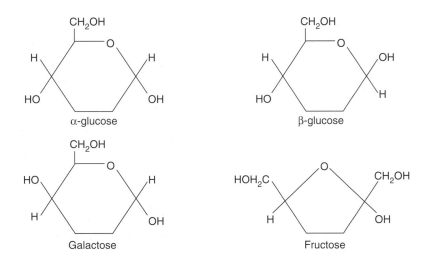

Hexose sugars like α-glucose are the monomers from which many other carbohydrates are formed. Two α-glucose molecules may be joined by condensation to form a molecule of the disaccharide maltose. The bond forms between carbon 1 of one α-glucose molecule and carbon 4 of the other and is called a **glycosidic bond** (see Figure 3.6).

Q 3 Maltose is formed by the condensation of two α-glucose molecules. What is the molecular formula of maltose?

In a similar way, other disaccharides can be formed. Lactose, for example, the sugar found in milk, is formed from α-glucose and galactose while sucrose is formed from α-glucose and fructose.

When sugars such as α-glucose are boiled with Benedict's solution, an orange precipitate is formed as Cu(II) ions in Benedict's solution are reduced to Cu(I) ions. Because the arrangement of chemical groups in the sugar molecule enables it to reduce the Benedict's solution it is known as a **reducing sugar**. Glucose, fructose, maltose and lactose are all reducing sugars but sucrose is not. It is a **non-reducing sugar** and will only produce a positive test with Benedict's solution if it is first hydrolysed by boiling it with dilute acid. This splits the sucrose molecules into glucose and fructose, which are reducing sugars.

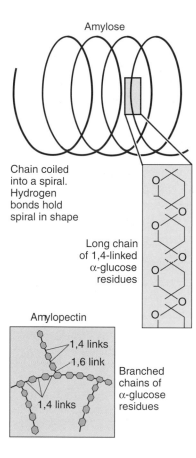

Amylose

Chain coiled into a spiral. Hydrogen bonds hold spiral in shape

Long chain of 1,4-linked α-glucose residues

Amylopectin

1,4 links
1,6 link
1,4 links

Branched chains of α-glucose residues

Figure 3.7
Starch consists of amylose and amylopectin.

Figure 3.8
Starch is a very important storage substance in plants. Starch stores such as those found in these barley grains form a vital energy source for most humans.

Starch and other polysaccharides

Starch is a mixture of two compounds, amylose and amylopectin. Both of these molecules are polymers containing a large but variable number of α-glucose molecules linked to each other by condensation.

Q 4 **Explain why the starch found in different plants may vary.**

Amylose consists of long straight chains of α-glucose molecules linked by 1,4-glycosidic bonds (see Figure 3.7). These chains are coiled to form a spiral, each coil in the spiral containing about six glucose residues. The coils of the spiral are held in place with hydrogen bonds. **Amylopectin** is also a polymer of α-glucose but its molecules are branched. This is because some of the glucose residues are joined by 1,6-glycosidic bonds.

Starch has a number of properties that make it an efficient storage molecule (see Figure 3.8), for example:

- molecules of both amylose and amylopectin are compact. This means that a lot of starch can be stored in a relatively small space

- it is easily broken down to glucose when needed for respiration

- starch is insoluble so it is more likely to remain in one place. In addition, as it is insoluble, it will not affect the water potential of the cells in which it is stored and bring about the movement of large amounts of water by osmosis.

Extension box 1

The main source of energy in the human diet is starch. In the UK, starch makes up about 30% of what we eat. The story of starch digestion, however, is not a simple one. Although all starch can be hydrolysed by the amylase enzymes found in the gut, there are many factors that determine the rate at which it is broken down. The starch found in processed foods, such as many breakfast cereals, is known as rapidly digestible starch. As its name suggests, it is broken down readily. Starch found in unripe bananas and potatoes is not broken down so readily, mainly because it is located in granules. This type of starch is known as resistant starch. It often travels the length of the small intestine and enters the large intestine or colon without being digested.

Figure 3.9
When the incidence of colon cancer is plotted against the daily intake of starch, the graph shows a negative correlation; the lower the starch intake, the higher the risk of colon cancer.

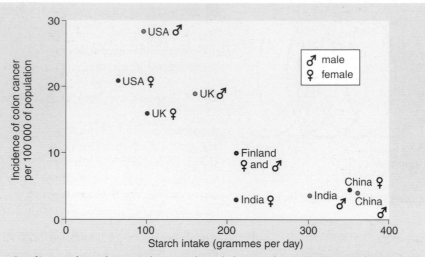

Figure 3.10
Eating a banana a day might be a good idea but it would have to be a green one that contains substantial amounts of resistant starch. In ripe bananas, those whose skins are yellow with black spots, most of the starch has been converted to sugars.

Studies such as the one that produced the results in Figure 3.9 have shown that there are clear links between the amount of starch in the diet and the incidence of colon cancer. Scientists have suggested that this link may be explained by the amount of resistant starch in the diet. As resistant starch enters the colon, it is broken down by the bacteria living there. These bacteria produce small fatty acid molecules such as butyric acid from digestion of the starch. Bacterial action may be helpful in two ways. The bacteria have a plentiful food source so they multiply rapidly. This helps to increase the rate of movement of faeces through the colon. Potential cancer-causing substances in the faeces therefore do not spend as long in contact with the cells that line the colon. Secondly, laboratory experiments have shown that butyric acid is very effective at preventing the growth of cancer cells.

Q 5 For most countries, the figures for starch intake in males are higher than those for starch intake in females. Suggest why.

Glycogen is another important polysaccharide. It is the main storage carbohydrate in humans. During a period of strenuous exercise, the body's glucose supplies are rapidly used up. Supplies can be replenished by breaking down glycogen stored in the liver and in the muscles. The molecules of glycogen are very similar to those of amylopectin but they are even more branched.

Cellulose is another polysaccharide. Its function is structural and it forms a very important component of plant cell walls. The monomer that forms cellulose is β-glucose. If you look back at Figure 3.5 you will see that β-glucose differs from α-glucose in that the H and OH groups on carbon 1 are the other way up. The β-glucose molecules in cellulose are also linked by condensation (Figure 3.11). In order to get the OH groups in carbon atoms 1 and 4 in the right position for a 1,4-glycosidic bond to be formed, one of the β-glucose molecules has to be 'flipped over'. This results in cellulose consisting of long straight chains with alternate β-glucose residues flipped over.

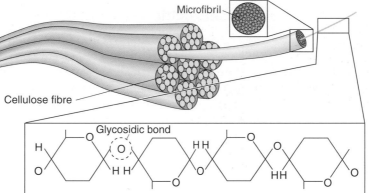

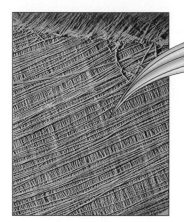

Microfibril

Cellulose fibre

Glycosidic bond

Long chain of 1,4 linked β-glucose residues

Figure 3.11
Cellulose and plant cell walls.

Cellulose molecules lie side by side and are linked to each other by hydrogen bonds to form bundles called **microfibrils**. The microfibrils are in turn held together in fibres. A cell wall is built up of many of these fibres running in different directions. Cellulose fibres have a high **tensile strength**. This means that they can withstand a very large pulling force without breaking and this helps to give the plant cell wall its strength.

Proteins

Each of the substances we considered in the previous section was built up from a single type of monomer. Not surprisingly, there are relatively few different sorts of polysaccharide because there are relatively few types of carbohydrate monomer. Proteins are very different. The basic building blocks of proteins are amino acids. Because there are 20 different amino acids found in proteins and because they can be put together in any order, there is a huge number of different proteins and they have many functions in a living organism (see Figure 3.12).

Figure 3.12
The body of a mammal contains more protein than any other organic compound. These proteins have a variety of different functions. Extension box 2 looks at how the functions of proteins are related to their structure.

Q   **6** A bacterial cell contains 1050 different types of protein but only one sort of polysaccharide. Explain why.

Amino acids

The 20 different amino acids that occur naturally and make up proteins have the same general structure. There is a central carbon atom, called the α-carbon, to which four groups of atoms are attached. These are an amino group ($-NH_2$), a carboxylic acid group ($-COOH$), a hydrogen atom ($-H$) and another group often referred to as the R group. The first three groups are always the same. The R group differs from amino acid to amino acid. Figure 3.13 shows the general structure of an amino acid. It also shows the molecular structures of three particular amino acids found in human proteins.

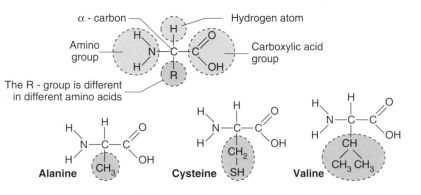

Figure 3.13
The structure of amino acids.

Figure 3.14
Joining amino acids.

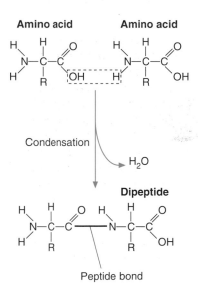

Amino acids can be joined together by condensation. A hydrogen atom is removed from the amino group of one amino acid and this combines with an $-OH$ group removed from the carboxylic acid of the other, forming a molecule of water. The bond formed between the two amino acid residues is known as a **peptide bond**. Joining two amino acids together produces a **dipeptide**. When a small number of amino acids are joined in this way, we have a **peptide**. Longer chains are called **polypeptides**. As with other biologically important polymers, polypeptides can be broken down to their constituent amino acids by hydrolysis. Figure 3.14 shows how a dipeptide can be formed from two amino acids.

Q 7 A peptide contains nine amino acids. How many peptide bonds will there be in a molecule of this peptide?

Polypeptides and proteins

We can join more amino acids together in this way to give a polypeptide chain. A **protein** consists of one or more of these polypeptide chains folded into a complex three-dimensional shape. Different proteins have different shapes. These shapes are determined by the order in which the amino acids are arranged in the polypeptide chains. This is the **primary structure** of the protein. Genes carry the genetic code that enables cells to make polypeptides and ensures that the sequence of amino acids is the same in all molecules of a particular polypeptide. Changing a single one of these amino acids may be enough to lead to a change in the shape of the protein and prevent it from carrying out its normal function. Figure 3.15 shows the primary structure of an enzyme called ribonuclease,

Q 8 Use Figure 3.14 to explain the meanings of the terms *N-terminus* and *C-terminus* for the ribonuclease molecule shown in Figure 3.15.

Figure 3.15
The primary structure of a protein, the enzyme ribonuclease. The names of the amino acids have been abbreviated.

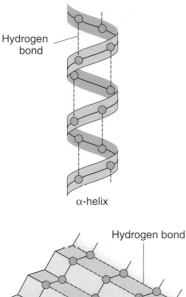

Hydrogen bond

α-helix

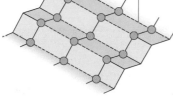

Hydrogen bond

β-pleated sheet

Figure 3.15
The primary structure of a protein, the enzyme ribonuclease. The names of the amino acids have been abbreviated.

Figure 3.16
There are two main types of secondary structure in a protein. In an α-helix, the polypeptide chain is coiled. In a β-pleated-sheet, the polypeptide is folded.

which breaks down, or hydrolyses, ribonucleic acid (RNA). It is made up of 124 amino acids linked together with peptide bonds.

The polypeptide chain may form a particular shape, which is held in place by hydrogen bonds between the amino acid residues. This produces the **secondary structure** of the protein (Figure 3.16). Sometimes the chain of amino acids, or part of it, may coil to form a spiral known as an α-helix. In other proteins, a β-pleated sheet is formed. This occurs when two parts of the chain lie parallel to each other and hydrogen bonds link one part of the chain to the other. Whether or not an α-helix or a β-pleated sheet is formed depends on the sequence of amino acids in the polypeptide. Some sequences are more likely to form an α-helix while others form β-pleated sheets.

The secondary structure involves twisting or folding of parts of the polypeptide chain. The whole chain may be further folded to give the protein molecule a complex globular shape known as its **tertiary structure**. As with the secondary structure, the tertiary structure of a protein is determined by the sequence of amino acids in the polypeptide chain. Since all molecules of a particular protein, such as ribonuclease, have the same sequence of amino acids, they will always fold in the same way to produce molecules with the same three-dimensional shape. This shape is extremely important and is very closely related to the function of the protein (see Extension box 2). Different types of bond form between different amino acids and help to maintain the shape of the protein. These bonds include:

- hydrogen bonds, which are formed between the R groups of a variety of different amino acids. These bonds are easily broken but very numerous

- disulphide bonds formed between residues of the sulphur-containing amino acid cysteine. These are fairly strong bonds which are particularly important in structural proteins such as those found in skin and hair.

Q 9 In which of the following are hydrogen bonds present
   (a) maltose
   (b) starch
   (c) the primary structure of a protein
   (d) the secondary structure of a protein?

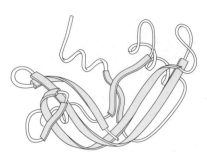

Figure 3.17 (above)
This diagram shows both the secondary and tertiary structure of a molecule of ribonuclease. The three spiral portions of the polypeptide chain represent the parts which are coiled into an α-helix. The flat pieces show where the polypeptide chain is folded into a β-pleated sheet.

Figure 3.18 (top right)
Antibody molecules are produced by the body in response to a foreign molecule or cell, such as that of an invading bacterium. Antibodies are proteins that have a quaternary structure. This model shows the two long polypeptide chains and the two shorter ones that make up an antibody molecule.

Figure 3.19 (right)
Haemoglobin is the red oxygen-carrying pigment found in the blood of humans and many other animals. It is an example of a globular protein. Collagen, on the other hand, is a fibrous protein. It is found in many parts of the human body, including the skin, cartilage and the walls of larger blood vessels.

Heating a protein causes the bonds that maintain its tertiary structure to break. The protein is said to be **denatured**. As a result, the molecules lose their shape and can no longer carry out their function (see Chapter 4). Figure 3.17 shows the secondary and tertiary structures of the ribonuclease molecule, whose primary structure is illustrated in Figure 3.15.

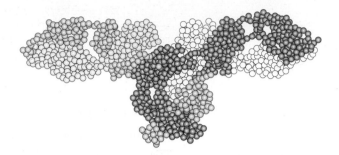

Some proteins are made up from more than one polypeptide chain. The way in which two or more polypeptide chains combine gives the protein its **quaternary structure** (see Figure 3.18). The chains are held together by the same sorts of chemical bond that maintain the tertiary structure. The enzyme ribonuclease does not have a quaternary structure because it consists of only one polypeptide chain. Each molecule of haemoglobin is made up of four polypeptide chains so it can be described as having a quaternary structure.

Haemoglobin
One of the four polypeptide chains that make up a molecule of haemoglobin

Collagen
Each polypeptide chain in a molecule of collagen is coiled into a helix

A haem group:
Each polypeptide chain is attached to a haem group. This group is important in transporting oxygen

A collagen molecule consists of three polypeptide chains coiled round each other

Protein molecules have one of two basic shapes: they may be curled up into a ball, in which case they are known as **globular proteins**, or they may be long and thin, and known as **fibrous proteins**. Globular proteins are usually soluble and play an important role in the metabolism of living organisms. Examples of globular proteins are the carrier proteins found in plasma membranes, enzymes and haemoglobin (see Figure 3.19). Fibrous proteins, on the other hand, as their name suggests, form long chains. They are insoluble and have important structural functions.

Extension box 2

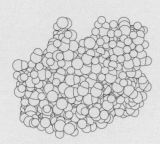

Figure 3.20
This model shows the three-dimensional shape of the enzyme ribonuclease. This is a protein whose function depends on the shape that results from its tertiary structure.

Figure 3.21
Diagram A represents a molecule of ribonuclease. The dots represent cysteine, an amino acid that contains sulphur. Disulphide bonds form between these amino acids and maintain the tertiary structure and the shape of the enzyme. In diagram B, the enzyme has been treated with mercaptoethanol. The bonds between the cysteine residues have broken. The ribonuclease molecule has unravelled and lost its tertiary structure. It will no longer function.

Figure 3.22
This person is receiving her yearly influenza injection, which involves giving the individual concerned a vaccine containing harmless, inactive influenza viruses. These stimulate some of her white cells to produce influenza antibodies if she is exposed to living influenza viruses. A different vaccine has to be used each year. This is because there are many different strains of influenza virus. An antibody that recognises one strain may not recognise another. Once again, it is a matter of the shape of a protein, this time an antibody, not matching and therefore fitting the shape of another found on the surface of the influenza virus.

Locks, keys and proteins

Globular proteins have many functions in living organisms. Their molecules have a tertiary structure that gives them a distinctive shape and this shape is related to their function (see Figure 3.20).

You will come across many globular proteins during your A-level course. They have different functions but the basic principle concerning the way in which they work is always the same. Each protein has a unique sequence of amino acids that gives it a particular tertiary structure and a distinctive shape. This shape allows other molecules to fit specific sites on its surface, and the protein to carry out its function.

Proteins as enzymes

We have seen that ribonuclease is an enzyme that breaks down the nucleic acid RNA into smaller components. It is a globular protein so it has a distinctive tertiary structure. Like all enzymes, it functions because part of its molecule forms what is called an active site. RNA molecules will fit into this site where they are broken down into their components. The tertiary structure of a protein is held by chemical bonds between the amino acids. One type of bond is the disulphide bond, which is formed between amino acids that contain sulphur. When ribonuclease is incubated with mercaptoethanol, these bonds are broken and the enzyme loses its tertiary shape. It no longer works. This is shown in Figure 3.21. Enzymes are described in detail in Chapter 4.

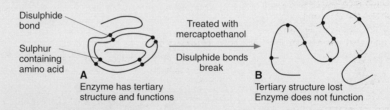

Proteins as antibodies

The production of antibodies is one of the main ways in which the white cells in the body respond to viruses and bacteria. Antibodies recognise these harmful microorganisms and attach to particular molecules on their surfaces. This triggers off mechanisms that lead to their destruction and the infection is controlled (see Figure 3.22). Antibodies, however, are very specific. One that is effective against a chickenpox virus, for example, will not recognise a measles virus. Once again, it is all down to shape (see Figure 3.23).

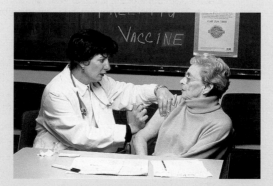

Figure 3.23
These diagrams show changes in the shapes of the protein coats of a particular strain of influenza virus between 1968 and 1985. The red-coloured areas are spikes on the protein coat of the virus.

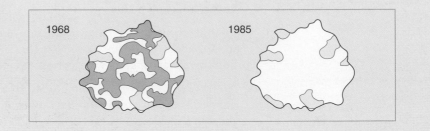

Q 10 Explain why a vaccine that was effective against the strain of influenza virus found in 1968 would not be effective against the strain found in 1978.

Proteins in the nervous system

Nerve cells or neurones do not join directly to one another. There are small gaps between them called synapses (see Figure 3.24). An impulse is transmitted along a nerve until it reaches a synapse. It then causes molecules of a neurotransmitter substance to be released. The neurotransmitter molecules diffuse across the gap and fit into protein receptor molecules. This results in a nerve impulse being triggered in the second neurone.

Figure 3.24
This electronmicrograph shows a synapse between two nerve cells. The round structures are vesicles containing neurotransmitter molecules.

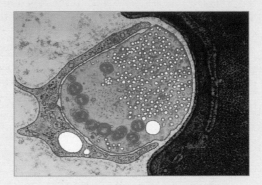

There are many different neurotransmitters. An important neurotransmitter in synapses in the brain is a substance called serotonin (Figure 3.25). Serotonin has molecules that have very similar shapes to those of a number of drugs that affect the brain. The similar shapes of these drug molecules means that they can all fit into the proteins that form the serotonin receptors.

Figure 3.25
The molecules of serotonin and LSD, drugs that affect the brain. Notice how they are very similar in shape.

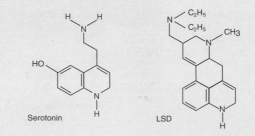

Lipids

Triglycerides

Figure 3.26
The body shape of this seal is due to a thick layer of fat under its skin. This insulates it from the cold water in which it lives. Fats also act as an important store of energy.

key term

The name **lipid** is used to describe a range of substances. Some of the most important of these are the **triglycerides**, usually known as fats and oils.

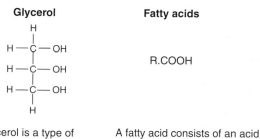

Glycerol

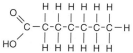

Fatty acids

R.COOH

Glycerol is a type of alcohol. It has three OH groups each of which can combine with a fatty acid

A fatty acid consists of an acid COOH group and a long hydrocarbon chain consisting of carbon and hydrogen atoms. This is represented by the letter R.

Saturated fatty acid

Unsaturated fatty acid

With the exception of the last one, each of the carbon atoms in the hydrocarbon chain of a saturated fatty acid is joined to two hydrogen atoms

In unsaturated fatty acids, there are double bonds between some of the carbon atoms. The hydrogen chain is not saturated with hydrogen atoms

Figure 3.27
Glycerol and fatty acids.

A triglyceride consists of three fatty acid molecules joined to a molecule of glycerol. Each **fatty acid** consists of an acid COOH group joined to a long hydrocarbon tail consisting of carbon and hydrogen atoms. The length of this hydrocarbon tail varies but in many of the fatty acids found in triglycerides there are between 14 and 16 carbon atoms. Some fatty acids have one or more double bonds between the carbon atoms in the tail. These are **unsaturated** fatty acids, so-called because they are unsaturated with hydrogen atoms (see Figure 3.27). The presence of double bonds produce kinks in the hydrocarbon chains. This stops the hydrocarbon chains of neighbouring triglycerides lying too close together and makes the lipid more fluid, lowering its melting point. As a consequence triglycerides that contain unsaturated fatty acids are often liquid at room temperatures. They are known as oils. Fatty acids that do not contain double bonds in their hydrocarbon chains are described as **saturated**. They usually have a higher melting point than unsaturated fatty acids and give rise to fats that are usually solid at room temperatures. Oils are more commonly found in plants while fats are usually associated with animals.

Q 11 **Explain what is meant by a polyunsaturated fatty acid.**

Glycerol is a type of alcohol. Its chemical structure is shown in Figure 3.27. Each of the three carbon atoms in a molecule of glycerol is associated with an OH group that is able to combine with a fatty acid (Figure 3.28). This reaction is a condensation, so when a triglyceride is formed, three molecules of water are also produced.

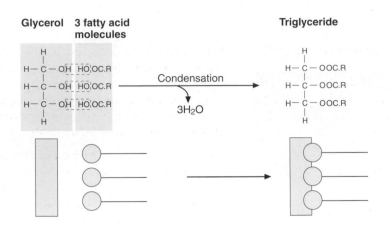

Figure 3.28
The formation of a triglyceride from a molecule of glycerol and three fatty acid molecules.

Phospholipids

Think of a phospholipid as having a 'head' consisting of the glycerol and phosphate and a 'tail' containing the long hydrocarbon chains of the two fatty acids (see Figure 3.29). The presence of the phosphate group means that the charge on the head of the molecule is unevenly distributed. It is said to be **polar** and is attracted to water. The head end of the molecule is described, therefore, as being **hydrophilic** or 'water-loving'.

The hydrocarbon tails do not have this uneven charge distribution. They are therefore **non-polar** and will not mix with water. The tail end of the molecule is described as being **hydrophobic** or 'water-hating'. This property means that if phospholipids are placed in water they will arrange themselves in a double layer with their hydrophobic tails pointing inwards and their hydrophilic heads pointing outwards. This double layer is called a **phospholipid bilayer** and it forms the basis of cell membranes (see Chapter 2).

Q 12 How does a phospholipid differ from a triglyceride?

key term

Phospholipids are very similar to triglycerides except that one of the fatty acids is replaced with a phosphate group.

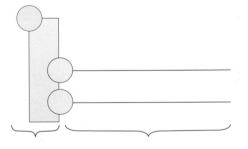

Head end of molecule. This is attracted to water and is described as **hydrophilic**

Hydrocarbon tail. This end of the molecule does not mix with water. It is described as **hydrophobic**

Figure 3.29
A phospholipid molecule. Because the head end of the molecule and the hydrocarbon tail have different properties, phospholipids arrange themselves in bilayers when placed in water.

Extension box 3

Figure 3.30 (above)
Oilseed rape has become increasingly popular as a crop in the UK over the past 25 years. Its bright yellow flowers are a familiar sight in spring. After the petals have withered and fallen from the flowers, the pods containing the seeds ripen and are harvested. Oil is extracted from the seeds.

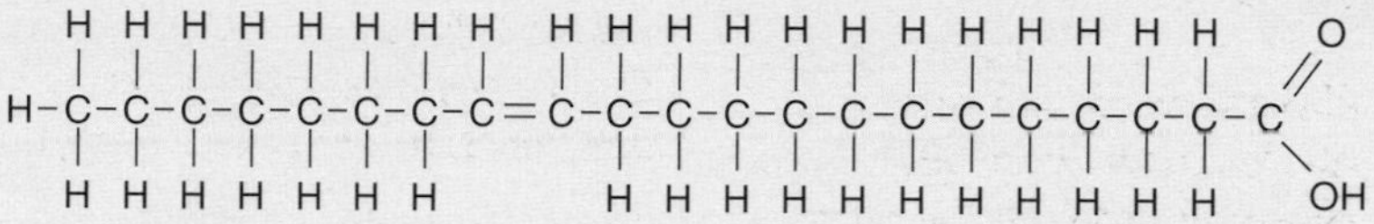

Figure 3.31 (right)
A molecule of erucic acid.

Fats and oils are often found in seeds. A large amount of energy is released when a triglyceride is respired, so triglycerides act as energy stores for the young plant as the seed germinates. Plants whose seeds store triglycerides are often important agricultural crops (see Figure 3.30). Every year almost 80 million tonnes of fats and oils are produced world-wide from this source.

There are many different varieties of oilseed rape and the oils that the seeds contain vary slightly in chemical composition. These differences are due to the fatty acids in the triglycerides of the seeds. In the 1970s, it was found that the presence in the diet of one of the fatty acids in rape seed oil, erucic acid (Figure 3.31), was linked with the accumulation of fat in the heart muscle of young animals. It was thought that if this happened in humans, it could lead to serious heath problems, so plant breeders set about breeding new strains of oilseed rape that were low in erucic acid.

Q 13 What type of fatty acid is erucic acid? Is it saturated, monounsaturated or polyunsaturated?

Varieties of oilseed rape low in erucic acid are now grown for use in the food industry. Oil from these plants is used in the production of margarine and cooking oil. Varieties high in erucic acid are still grown. The oil from these plants is used to produce erucimide, a substance that is used for coating plastics such as credit cards and plastic bags to prevent them becoming sticky.

Separating and identifying molecules using chromatography

Biochemists use chromatography to separate and identify the various substances present in a mixture. Paper chromatography uses a strip or square of absorbent paper (see Figure 3.33) but there are other sorts of chromatography that rely on different techniques. The substances to be separated are extracted and dissolved in a suitable solvent. A pencil line is ruled near the bottom of the paper. This forms the **origin**. A fine pipette or capillary tube is used to put drops of the solvent containing the mixture on the origin. A very small drop is placed on the paper, it is allowed to dry and another spot is put on top of the first. This process is repeated a number of times, producing a tiny concentrated spot.

Figure 3.32
In autumn, the leaves fall from many woody plants. A layer of cork forms at the bottom of the leaf stalk and this traps some of the sugars produced in leaves during photosynthesis. These sugars are converted to anthocyanins, which give leaves of plants such as this Virginia creeper their characteristic red colour. Leaf pigments are just one group of substances that can be investigated with chromatography.

The chromatography paper is now suspended in a jar containing a little solvent. The paper is arranged so that the bottom of it just dips into the solvent. The lid is placed on the jar so that the air inside becomes saturated with solvent vapour and the apparatus is left on one side. The solvent moves up the paper. The different substances in the mixture also move but they move at different rates and are separated out. When the solvent has nearly reached the top of the paper, the paper is removed and the position of the **solvent front** is marked with a pencil.

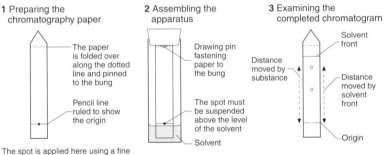

Figure 3.33
Carrying out paper chromatography.

If you are separating coloured substances such as the pigments in a leaf, you will see different coloured areas on the paper, each one representing a different pigment. Many substances are not coloured, however, and the chromatogram has to be developed before the positions of the individual substances can be seen. Developing is usually done by spraying the chromatogram or dipping it in an indicator that changes colour when it is mixed with the substance concerned.

The process described so far allows different substances to be separated but it does not identify them. To do this we calculate the **Rf value**. This is a measure of how far the substance has moved compared with the distance moved by the solvent. It can be calculated from a simple equation:

$$Rf\ value = \frac{distance\ moved\ by\ substance}{distance\ moved\ by\ solvent\ front}$$

Each compound in the mixture has a unique Rf value so we can use this to identify which spot represents which substance.

Q 14 **Why does the Rf value of a substance always have a value less than 1?**

Sometimes the spots containing the substances do not separate very much. When the chromatogram is removed some of the spots are very close together. In situations like this we can use two-way chromatography (Figure 3.34). A square piece of paper is used. When the solvent front is near the top of the paper, the chromatogram is removed and turned through 90° so that the spots are at the bottom of the paper. The paper is then suspended in a different solvent, which is again allowed to run up the paper.

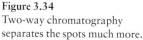
Figure 3.34
Two-way chromatography separates the spots much more.

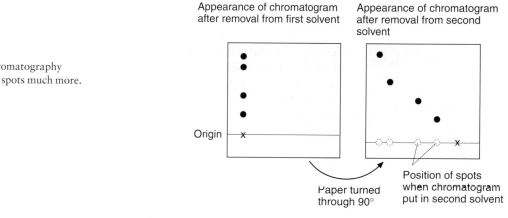

Appearance of chromatogram after removal from first solvent

Appearance of chromatogram after removal from second solvent

Origin

Paper turned through 90°

Position of spots when chromatogram put in second solvent

Summary

- Biological molecules are based on a small number of chemical elements and frequently consist of monomers joined by condensation into polymers. These polymers may be broken down by hydrolysis to give the monomers from which they are built.

- Starch and glycogen are formed by the linking of α-glucose molecules with the formation of glycosidic bonds.

- Cellulose is formed by the linking of β-glucose molecules.

- Amino acids may be linked together with peptide bonds to produce dipeptides and polypeptides. Polypeptides form proteins which show different levels of structure.

- A molecule of glycerol and three molecules of fatty acids combine to produce a triglyceride.

- Chromatography can be used to separate and identify molecules in a mixture.

Assignment

Figure 3.35
The ability to feed their young on milk is a characteristic of all mammals.

This exercise is designed to introduce you to the skills involved in interpreting data. The first two questions will help you to understand the data presented in the table. You will then have to change or translate the information in the table into another form, a bar chart. The final part of the exercise involves interpretation of information.

All mammals share certain characteristics. They have hair on their body surface and have sweat glands that help in controlling body temperature. They also feed their young on milk produced from the mammary glands of the female (see Figure 3.35).

Milk must provide a young mammal with the nutrients it requires before it is old enough to eat food similar to that of an adult. In addition to protein, carbohydrate and lipid, milk also contains other substances such as vitamins and mineral ions. The relative amounts of these substances, however, differ from one species of mammal to another. Table 3.2 shows the composition of human milk and seal milk.

Species	Mass of substance (g per 100 g of milk)			
	Protein	Lactose	Lipid	Ash
Human	1.2–1.5	7.0	3.8	0.2
Harp seal	13.8	0	36.5	0.6

Table 3.2

1 Explain why the figures are given per 100 g of milk.

(2 marks)

2 All the water can be removed from a milk sample and the dry material that is left can then be heated strongly and burnt. What remains after burning is called ash.

(a) Explain why there will be no carbon-containing substances in the ash.

(2 marks)

(b) Use the information in the question to suggest why it is useful to measure the amount of ash obtained from the sample.

(1 mark)

3 Draw a bar chart to compare the mass of these substances in human and harp seal milk.

(4 marks)

Harp seals live in the Arctic. They give birth to their young on large areas of ice. These areas have few predators but there disadvantages for the young seal because there is very little shelter and it is extremely cold. When it is born, a typical young harp seal weighs 10.8 kg. After 9 days of feeding on its mother's milk, its body mass will have increased to 34.4 kg.

4 (a) Calculate the percentage increase in mass of the young harp seal over this 9 day period. Show your working.

(2 marks)

(b) Explain how the increase in mass of the young harp seal can be related to the composition of the milk on which it feeds.

(2 marks)

The composition of the milk that is produced depends to a certain extent on the food that is eaten. In a study of human breast milk, samples were collected from two groups of women. Those in one group were vegans and only ate food obtained from plants. Those in the other group, the control group, ate food obtained from both animals and plants. Table 3.3 shows the concentrations of different fatty acids in the milk samples.

Fatty acid	Number of double bonds in hydrocarbon chain	Number of carbon atoms in hydrocarbon chain	Concentration of fatty acid in mg per g of milk	
			Vegan group	Control group
Lauric	0	12	39	33
Myristic	0	14	68	80
Palmitic	0	16	166	276
Stearic	0	18	52	108
Palmitoleic	1	16	12	36
Oleic	1	18	313	353
Linoleic	2	18	317	69
Linolenic	3	18	15	8

Table 3.3

5 (a) Explain why the first four fatty acids in this table may be described as saturated.

(1 mark)

(b) Calculate the total concentration of saturated and unsaturated fatty acids from milk in the vegan group and from milk in the control group. Write your answers in a suitable table.

(2 marks)

(c) Describe and explain the main differences between the occurrence of saturated and unsaturated fatty acids in milk produced by the vegan group and the control group.

(3 marks)

Examination questions

1 Table 3.4 shows the number of carbon atoms contained in some substances.

Substance	Number of carbon atoms
Amino acid	2–11
Glycerol	
Glucose	
Starch	Large and variable number

Table 3.4

(a) Complete the table to show the number of carbon atoms in glycerol.

(1 mark)

(b) Explain why the number of carbon atoms

(i) may vary from 2 to 11 in an amino acid molecule;

(1 mark)

(ii) may be described as variable in a starch molecule.

(1 mark)

(c) Explain why the type of chemical reaction in which glucose is converted to starch may be described as condensation.

(2 marks)

2 Describe how the process of paper chromatography is carried out in order to separate substances in a mixture.

(3 marks)

(b) Some fatty acids were synthesised in which all the carbon atoms were radioactive. A mixture of three of these fatty acids was separated using paper chromatography. The level of radioactivity in each of the fatty acids was measured by passing the chromatography strip under a Geiger counter and the level was plotted as a graph. The results are shown in the Figure 3.36.

(i) The *total* level of radioactivity in each of the fatty acids was measured by estimating the total area under each curve rather than the height of each peak. Explain why.

(1 mark)

(ii) Suggest **one** reason why the three fatty acids showed different levels of radioactivity

(1 mark)

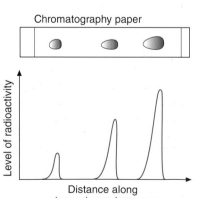

Chromatography paper

Level of radioactivity

Distance along chromatography paper

Figure 3.36

Enzymes

The greenish-yellow glow of a firefly on a summer night in Malaysia is produced by one of the most interesting enzymes ever discovered. It is called luciferase and it catalyses the breakdown of a protein called luciferin. During this reaction, most of the energy is released as light rather than heat, causing the glow we can see in the dark. Colonies of bacteria that contain a similar light-releasing enzyme inhabit four species of 'flashlight fish', deep-sea fish with glowing pockets directly beneath their eyes (Figure 4.1). Each fish provides nourishment for the millions of bacteria in its eye pockets; in return, the luminous bacteria help the fish prey and mate in the inky reaches of the deep ocean.

Luciferase acts on luciferin only if energy is present in the form of adenosine triphosphate (ATP):

$$\text{luciferin} \xrightarrow[\text{luciferase}]{\text{ATP}} \text{product} + \text{light} + \text{heat}$$

Researchers have been using this 'glowing enzyme' to detect how much ATP is present in certain systems. For example, luciferase and luciferin can be added to blood stored in blood banks. If the red blood cells are stored too long they begin to degenerate and leak ATP molecules into surrounding fluid. These molecules will in turn drive the breakdown of luciferin by luciferase and the blood will glow in the dark. Bacteria also release ATP and the same method is used to determine if they are present in urine so that infections may be detected and treated.

Luciferase and similar enzymes are truly illuminating tools for biologists, as well as providing headlights for fish and tail-lights for fireflies.

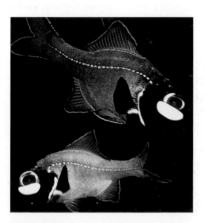

Figure 4.1
Flashlight fish.

Enzymes are amazing molecules. They are produced by living cells, with each cell containing several hundred enzymes. They are extremely specific, generally reacting with only one substrate. They can speed up the rate of chemical reactions by as much as a million times and at the end of the reaction they remain unchanged. Within a cell, in the absence of enzymes reactions would take place at too slow a rate to sustain life. To increase the rate of reactions, high temperatures would be necessary and this would be lethal to a cell.

Q 1 By what process are enzymes made within the cell?

key term

Enzymes are globular proteins that catalyse chemical reactions in living organisms.

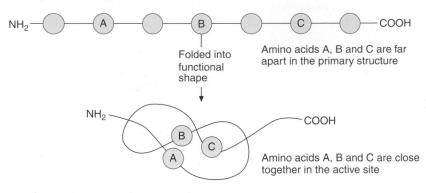

All enzymes are proteins

To understand how enzymes work it is necessary to understand the structure of proteins.

A protein is made up of a unique sequence of amino acids, known as its primary structure. However, the protein molecule does not lie flat: in a watery environment, such as inside living cells, the primary structure of the molecule will fold of its own accord into a unique and precise three-dimensional shape, held together by a number of different bonds (see Chapter 3).

The shape of the enzyme has to be just right: if the enzyme molecule is deformed, even slightly, it will not function.

Each different enzyme has its own highly specific shape, with a 'pocket' at a particular position. The pocket is known as the **active site**, and it is here that a substrate binds.

Only a few amino acids, normally between 3 and 12, in the chain of an enzyme are actually involved in the enzyme-controlled process. The specific amino acids that form the active site come together from different positions in the primary structure of the protein (see Figure 4.2). It is here that binding of the substrate occurs. The remaining amino acids make up the bulk of the enzyme, which maintains the correct globular shape of the molecule.

Activation energy

Consider a molecule of sucrose. It consists of a molecule of glucose and a molecule of fructose joined by a glycosidic bond. Sucrose can be hydrolysed into glucose and fructose by the addition of water to break the glycosidic bond (Figure 4.3).

key term

Activation energy is the energy needed to bring molecules together so that they will react with each other.

Figure 4.3
The reaction between sucrose and water will occur slowly due to random collisions.

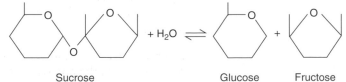

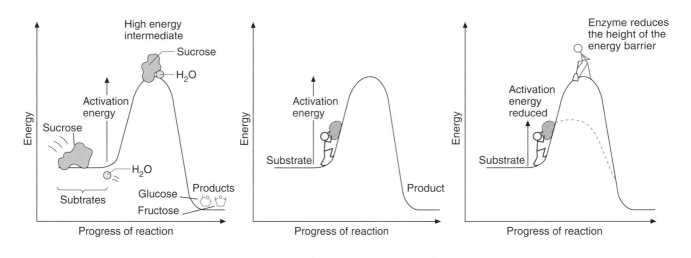

Figure 4.4
Energy diagram. Sucrose and water must collide with sufficient energy to form the unstable high-energy intermediate if they are to react to form glucose and fructose.

Figure 4.5
Energy diagram illustrating that energy needs to be added to a reaction to allow it to proceed.

Figure 4.6
Energy diagram illustrating that enzymes reduce the activation energy.

Without an enzyme, molecules react by randomly colliding, but energy has to be applied. For a reaction to take place molecules, known as **substrates,** have to collide with enough energy to break and form bonds, creating **products**. In this case, the sucrose and water collide to form glucose and fructose. This reaction could be carried out in a test-tube but it would occur very slowly.

The energy required to make substrates react is called the **activation energy**. This can be illustrated in an energy diagram (see Figure 4.4). At the start of the reaction, sucrose and water have a certain amount of energy. They collide with one another and form an unstable high-energy intermediate which quickly changes into glucose and fructose.

In rearranging the bonds of sucrose and water some energy is released, so the products have less energy than the initial molecules. The minimum amount of energy needed to start the reaction, leading to the formation of an unstable intermediate, is the activation energy.

Every chemical reaction thus has an energy barrier that has to be overcome before a reaction can occur. A comparison that is often used is that of a boulder resting on top of a hill (Figure 4.5). Although it will naturally roll down the hill the boulder is prevented from doing so by a small mound of earth.

There are two ways to get the boulder to roll down the hill. You can supply enough energy to push it to the top of the mound, where it can then roll down by itself. This is equivalent to supplying heat to start a reaction.

Alternatively, you could dig away at the mound, reducing the energy needed to push the boulder to the top of the mound. This is equivalent to supplying an enzyme to a reaction (Figure 4.6).

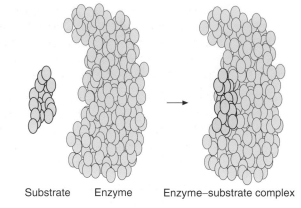

Figure 4.7
The binding of the enzyme ribonuclease with its substrate RNA, forming an enzyme–substrate complex.

Substrate   Enzyme   Enzyme–substrate complex

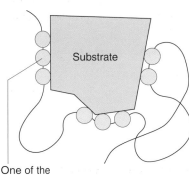

Substrate

One of the amino acids forming the active site

Figure 4.8
Diagram illustrating the position of the amino acids forming the active site.

Instead of actually supplying energy, the enzyme reduces the height of the energy barrier and therefore reduces the activation energy necessary for a reaction to take place. This may not seem very impressive but the involvement of an enzyme increases the rate of reaction by a billion times! Also, by lowering the activation energy, the reaction can take place at the temperatures and pressures which exist inside living cells.

Binding for activity

Substrate molecules bind to the active site of an enzyme. See for example the binding of the enzyme ribonuclease with its substrate RNA, forming an enzyme–substrate complex (Figure 4.7).

The substrate molecules are usually very much smaller than the enzymes. When substrates bind to the enzyme they slot neatly into the active site and are close to the amino acids which form it (Figure 4.8).

At first it was thought that an enzyme's active site was merely a negative impression of its substrate. This idea was called '**lock and key**' because the substrate seems to fit into the active site as a key fits into a lock (Figure 4.9).

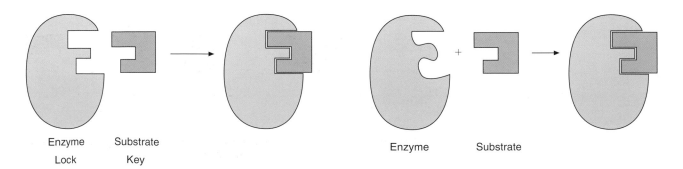

Enzyme   Substrate
Lock   Key

Enzyme   Substrate

Figure 4.9
The lock and key hypothesis.

Figure 4.10
The induced fit theory. The change in the active site brings the amino acids into their correct positions in the active site so a reaction can occur.

The enzyme was pictured as a rigid structure, which had an active site that was complementary to the substrate. However, this idea did not explain how other molecules could alter an enzyme's activity by binding to it at some other site that was not the active site. If the enzyme were flexible then such an effect could be explained.

When a substrate combines with an enzyme, it induces changes in the enzyme's shape. The amino acids which make up the active site are moulded into a precise form, which enables the enzyme to perform its catalytic function effectively. For this reason biologists refer to this modified theory as the **induced fit** of substrate to enzyme (Figure 4.10).

Thus the shape of the enzyme molecule is affected by the substrate, just as the shape of a glove is affected by the hand that is placed into it. The flexible enzyme molecule wraps itself around the substrate. The enzyme molecule in turn distorts the substrate in the resulting enzyme–substrate complex, causing a reaction to occur rapidly.

Once the reaction has occurred and products formed, which are a different shape to the substrate, they no longer bind to the active site and diffuse away. The flexible enzyme returns to its original shape, ready to bind to the next molecule of substrate.

Enzyme reactions

The enzyme combines reversibly with the substrate to form an enzyme–substrate complex:

$$\text{enzyme} + \text{substrate} \rightarrow \text{enzyme–substrate complex}$$
$$E + S \rightarrow ES$$

The enzyme–substrate complex then breaks down to give the product and releases the enzyme in an unchanged form:

$$\text{enzyme–substrate complex} \rightarrow \text{products} + \text{enzyme}$$
$$ES \rightarrow P + E$$

We can represent this using a simple diagram (Figure 4.11).

Figure 4.11
Illustration of an enzyme-controlled reaction.

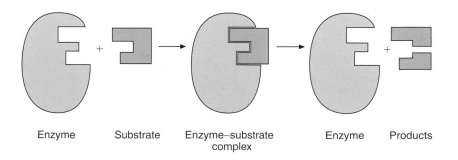

| Enzyme | Substrate | Enzyme–substrate complex | Enzyme | Products |

Enzyme controlled reactions are affected by a number of factors. How they affect the reactions can be understood using the principle described above.

Effect of temperature on enzyme activity

The graph in Figure 4.12 shows the rate of a typical enzyme-catalysed reaction and how it varies with temperature. The rate of an enzyme reaction is measured by the amount of substrate changed or the amount of product formed during a period of time.

At low temperature, say between 5°C and 30°C, increasing the temperature provides more heat energy. This increases the kinetic energy and makes both the enzyme molecules and the substrate molecules move faster. There will be an increase in the number of collisions between the substrate and the active site of the enzyme, resulting in more enzyme-substrate complexes and in turn the formation of more products.

As the temperature continues to increase, to above 40°C, the enzyme and substrate molecules move even faster. However, the structure of the enzyme molecule vibrates so energetically that some bonds holding the tertiary structure break. This is especially true of hydrogen bonds. The enzyme begins to lose its globular shape, which affects the active site such that the substrate will no longer fit into it. The enzyme is said to be denatured and will not regain its correct shape even if the temperature is lowered.

The temperature at which an enzyme catalyses a reaction at the maximum rate is called the optimum temperature. This can vary considerably for different enzymes, from 2°C to 78°C.

If the temperature is reduced to near or below freezing point, enzymes are inactivated, not denatured. They will regain their function when higher temperatures are restored.

Q 2 Why is it an advantage for humans to have a constant body temperature of 37°C.

3 Egg albumen, when heated goes solid and white. What do you think has happened to the proteins in the egg white and why does the white not become runny again when the egg is cooled?

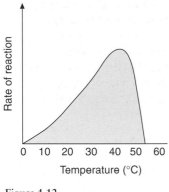

Figure 4.12
The effect of temperature on an enzyme-controlled reaction.

Extension box

Until recently, heat-loving, **thermophilic**, organisms e.g. the bacterium *Thermus aquaticus*, which grows optimally at 70°C in hot volcanic springs, were thought to exist at the highest temperatures possible for life.

In the last 15 years however, bacteria that grow in temperatures beyond the boiling point of water (100°C) have been found. *Pyrococcus furiosus* (the 'furious fireball') and *Thermotoga maritima* known as **hyperthermophiles** are not able to grow below 70°C.

Many biological molecules, which include enzymes, DNA and RNA are denatured rapidly by heat; the theoretical limit for stability was thought to be close to 90°C, the temperature above which the DNA helix cannot form. However, nature defies such predictions; the record for thermo-tolerance is held by *Pyrodictium abyssi*, which can grow at

113°C. DNA from hyperthermophiles is no different from that of other organisms. However, in all cells, DNA is usually wrapped in proteins known as histones, which protect it from breakage but allow contact with the enzymes that replicate the DNA. The stability of DNA in hyperthermophiles is probably due to an unusually heat resistant (thermostable) form of the histones. Proteins consist of chains of amino acids (polypeptides) folded in a very precise manner. Heat damages proteins by breaking hydrogen bonds that maintain the tertiary shape of the protein.

Proteins and enzymes in hyperthermophiles are remarkably thermostable. For example, a starch-degrading enzyme, α-glucosidase, from *Pyrococcus* is optimally active at 115°C. However, enzymes from organisms that live at normal temperatures and the same type of enzymes from hyperthermophiles appear structurally similar. When the structures of enzymes are compared, it is clear that their overall structures are remarkably similar, so the difference that allows the protein from the hyperthermophiles to survive must lie within the fine detail.

Several factors may account for the thermostability of proteins in hyperthermophiles.

- More bonds may be used to maintain the protein shape.

- The loops of polypeptides extending from the protein surface may be absent or reduced.

- There may be fewer of the amino acids that are particularly unstable at high temperature.

Figure 4.13
A hot spring in Yellowstone National Park, USA.

More than 80% of the biosphere, including many areas of the oceans, never exceeds 5°C, yet life abounds there. Even colder habitats exist in polar regions, for example the surface of ice floes where high concentrations of salt depress the freezing point of the water. Here grow the cold-loving organisms, **psychrophiles**. The lowest temperature at which life is known to exist is −10°C. Sub zero temperatures cause water to freeze in the cell, denying the cell free water molecules for metabolism. Ice crystals which form at these temperatures can rupture cell membranes. Most psychrophiles grow very slowly and might be expected to have specially adapted cell components. Indeed, proteins from *Bacillus TA41* isolated from Antarctic seawater are adapted to be active at low temperatures. Less is known about psychrophiles than about other extremophiles because they are so difficult and slow to grow. Also their biotechnological potential is less obvious.

Uses in biotechnology for the enzymes from extremophiles are emerging rapidly. Most enzymes are not robust molecules and industry often employs reactions that proceed at relatively high temperatures.

- The polymerase chain reaction (PCR), which enables many copies of specific, DNA sequences to be made in a test tube.

- The food processing industry uses a great many enzymes, such as proteases, lipases and cellulases.

- Many domestic detergents operate best in warm, alkaline conditions. Biological washing powders already contain enzymes from thermophiles.

Effect of pH on enzyme activity

pH is a measure of the concentration of hydrogen ions in a solution. The higher the hydrogen ion concentration, the lower the pH. Most enzymes function efficiently over a narrow pH range. A change in pH above or below this range reduces the rate of enzyme activity considerably (Figure 4.14).

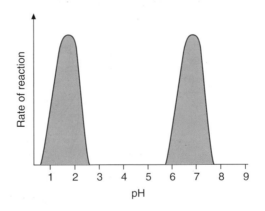

Figure 4.14
The effect of pH on the rate of an enzyme-controlled reaction. The diagram shows two enzyme-controlled reactions, each having a different optimum temperature.

pH has two main effects on the action of an enzyme:

- changes in pH lead to the breaking of the ionic bonds that hold the tertiary structure of the enzyme in place. The enzyme begins to lose its functional shape, particularly the shape of the active site, such that the substrate will no longer fit into it. The enzyme is said to be denatured

- changes in pH affect the charges on the amino acids within the active site such that the enzyme will not be able to form an enzyme–substrate complex.

The pH at which an enzyme catalyses a reaction at the maximum rate is called the optimum pH. This can vary considerably from pH 2 for pepsin to pH 9 for pancreatic lipase.

Q 4 Amylase is an enzyme found in the buccal cavity at pH 8. What happens to this enzyme when it finds its way into the stomach?

Effect of enzyme concentration

Provided that the substrate concentration is maintained at a high level, and other conditions such as pH and temperature are kept constant, the rate of reaction is directly proportional to the enzyme concentration (Figure 4.15). The reason for this is that if more enzymes are present then there are more active sites available for the substrate to slot into.

Effect of substrate concentration

As the substrate concentration increases, for a given enzyme concentration, then the rate of reaction also increases. As there are more substrate molecules present, an enzyme's active site can bind with more new substrates in a given time (Figure 4.16).

However, if the substrate concentration continues to increase, with a constant enzyme concentration, there comes a point where every enzyme's active site is forming enzyme–substrate complexes at its maximum rate. If more substrate is added, the enzyme simply cannot bind with the substrate any faster; substrate molecules are effectively queuing up for an active site to become vacant. Thus any extra substrate has to wait until the enzyme–substrate complex has dissociated into products and left the active site before it can complex with the enzyme. The enzyme is said to be working at its maximum possible turnover rate and the rate of reaction reaches a plateau. The maximum possible turnover rate is usually defined as the number of substrate molecules turned into product in one minute by one molecule of enzyme. Values range from less than 100 to many millions (see Table 4.1).

Enzyme	Turnover rate
Carbonic anhydrase	36 000 000
Catalase	5 000 000
Chymotrypsin	6 000
Lysozyme	60

Table 4.1

At high substrate levels, both enzyme concentration and the time it takes for dissociation of the enzyme–substrate complex molecule limit the rate of reaction.

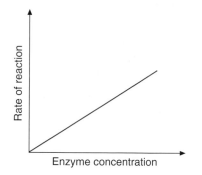

Figure 4.15
The effect of enzyme concentration on the rate of an enzyme-controlled reaction.

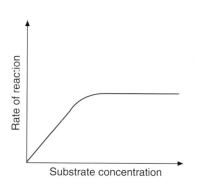

Figure 4.16
The effect of substrate concentration on the rate of an enzyme-controlled reaction.

Enzyme inhibition

Competitive inhibitors

These molecules have a shape similar to that of the enzyme's normal substrate so that they can fit into the active site and form an enzyme–inhibitor complex (Figure 4.17).

key term

Enzyme inhibitors are molecules that can reduce the rate of an enzyme-controlled reaction.

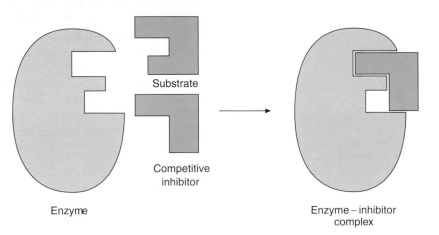

Figure 4.17
The competition between a competitive inhibitor and a substrate for the active site on an enzyme.

If the competitive inhibitor molecule is in the active site, no other molecule can enter the site. While it remains there it prevents access for any molecules of the true substrate.

The inhibitor and substrate compete for the active site. The molecule that is most likely to form a complex with the enzyme will be the one present in the highest concentration. If the concentration of substrate increases, the level of inhibition is reduced. So at high substrate concentrations the inhibitor has little or no effect because the relatively small number of inhibitor molecules is overwhelmed by the number of substrate molecules present (Figure 4.18).

Methanol poisoning is an example of competitive inhibition. Methanol (CH_3OH) can bind to the active site of the enzyme dehydrogenase, whose

Figure 4.18
The effect of substrate concentration on the rate of a reaction with and without a competitive inhibitor.

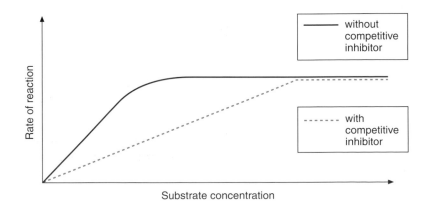

true substrate is ethanol (CH_3CH_2OH). A person who has accidentally swallowed methanol is treated by being given large doses of ethanol, which competes with methanol for the active site.

Non-competitive inhibitors

This type of inhibitor has no real structural similarity to the substrate and forms an enzyme–inhibitor complex in a 'pocket' on the enzyme other than its active site (Figure 4.19). It has the effect of altering the globular structure of the enzyme and the shape of the active site, so that even though the true substrate is present, it is unable to bind with the enzyme. In this case, if the concentration of the substrate is increased, the degree of inhibition is not affected because the increased number of substrate molecules does not affect the inhibitor's ability to bind with the enzyme. Even at high concentrations of substrate therefore a non-competitive inhibitor will always reduce the rate of reaction (Figure 4.20).

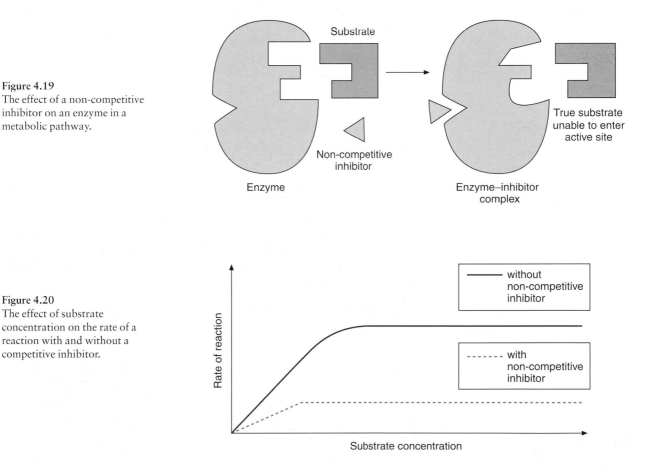

Figure 4.19
The effect of a non-competitive inhibitor on an enzyme in a metabolic pathway.

Figure 4.20
The effect of substrate concentration on the rate of a reaction with and without a competitive inhibitor.

Q 5 Write down the main differences between competitive and non-competitive inhibitors.

End-product inhibition

Metabolic reactions within a cell are normally multi-stepped reactions, with each step being controlled by a single enzyme (Figure 4.21). The end products of this process may accumulate within the cell and it may be important for the reaction to stop when sufficient product has been made. This is achieved by non-competitive inhibition of an enzyme earlier in the reaction sequence by the end product.

Figure 4.21
The effect of an end-product inhibitor on an enzyme earlier in a metabolic pathway.

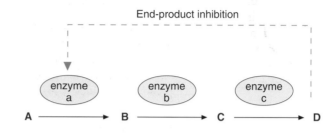

This is an example of a negative feedback mechanism serving to control an aspect of metabolic activity.

Figure 4.22
End-product inhibition. When sufficient isoleucine has been made, the isoleucine acts as a non-competitive inhibitor of threonine deaminase, changing the shape of the active site, thus preventing more threonine from reacting.

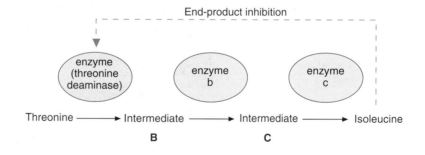

Summary

- Enzymes consist of globular proteins.

- Enzymes act as catalysts by lowering activation energy through the formation of enzyme–substrate complexes.

- There are two models of enzyme action: the lock and key and induced fit models.

- Temperature, pH, concentration of enzyme and concentration of substrate affect the rate of enzyme-controlled reactions.

- Competitive and non-competitive inhibitors decrease the rate of enzyme-controlled reaction.

Assignment

Most, if not all, biologists would describe their subject as a science and they would probably define biology as being the science of living organisms. But what is science? It is much more than the study of facts. Science is about solving problems by suggesting possible explanations or hypotheses and testing these with experiments. It is this that really defines science and makes it unique.

Biologists must obviously be able to design experiments, and you will be assessed on your ability to do this in your course work. Although every investigation is different, the principles adopted in designing an experiment are the same. In this assignment we will look at these basic principles and then apply them to one particular investigation.

Figure 4.23
Biology is different from other sciences you study for A-level in that you can carry out genuine research. It is quite likely that no-one has studied the plants in the area where this student is working.

We will start by taking an overall view. Designing and planning an experiment requires you to do three things.

- Make sure that you have a clear idea of what it is you are trying to test. Look at the simple sentence below. Filling in the gaps in this sentence should help you with this.

- As .... is changed, ... will change as a result.

- Identify the factor you need to change. This is the independent or manipulated variable. Decide how you will change it.

- Identify the factor that will change as a result. This is the dependent variable. You will need to decide how you will measure changes in the dependent variable so that you get quantitative results.

To summarise, make sure you know what it is that you are trying to do, decide how to change the independent variable and decide how to measure the dependent variable. Now, let us look at a particular problem.

When fruit like the tomatoes in Figure 4.24 ripen, several changes take place. The skin goes from green to red; they taste sweeter and they get softer. What causes them to get softer? We can use our knowledge of biology to suggest an explanation. Pectin is a substance found in and between plant cell walls which helps to bind cells together. As a fruit such as a tomato ripens, more of the enzyme pectinase is produced. Pectinase breaks down pectin into smaller, soluble molecules. As a result the cells separate from each other and the fruit gets softer.

Figure 4.24
Colour changes in ripening tomatoes are accompanied by changes inside the fruit which result in them becoming softer.

The first step is to go back to the incomplete sentence in the first bullet point and make sure that we know exactly what we are trying to do. Filling in the gaps gives us the idea that:

As *the softness of the fruit* is changed, *the amount of pectinase in the tomato* will change as a result.

What we need to do then is to collect some tomatoes, ranging from hard and firm to soft and squashy, and find a way of measuring their squashiness. You can't buy a squashometer so you will need to design one for yourself.

1 (a) Think about tomatoes. If you apply a constant force to tomatoes of different ripeness, a very ripe tomato would respond differently to an unripe tomato. How?

  (b) Use your answer to (a) to design a simple device that can be used to measure the squashiness of a tomato. You should describe it in sufficient detail for another member of your class to construct it without further help.

The next step is to isolate the pectinase enzyme from the tomatoes and measure its activity. You will need the information below to do this.

● Enzymes are soluble. The juice from a tomato will contain pectinase.

● Pectin is sold in many chemists and supermarkets. It is called Certo and is used by people who make jam because it helps the jam to set. Certo is a thick liquid. As it is broken down by pectinase it becomes thinner and flows more easily.

2 (a) Describe how you will get your samples of tomato juice from tomatoes of different softness.

  (b) Suggest how you could measure how easily a Certo solution flows.

  (c) Now draw a flowchart summarising the main steps in your procedure.

Another of the experimental skills you will need to develop is the ability to evaluate your ideas. You should be able to identify sources of error that are likely to affect the reliability of your results.

3 What are the steps in your procedure that are likely to make your results unreliable? Is there anything you can do about them at this stage?

Once we have the outline in place, we must ask ourselves if there are any other factors that could influence the results of this experiment and should therefore be kept constant. Remember, an enzyme is involved and there are a number of factors that influence the rate of enzyme-controlled reactions. If any of these are allowed to vary they will affect the reliability of the results.

4 List the factors that could influence the rate at which pectinase breaks down pectin. Suggest how you will keep each of these constant in your experiment.

There is a final experimentation step and it concerns setting up controls. The best approach to this is to look at the experiment and predict the results you would expect if it supports your hypothesis. In this case, we would expect the pectin to be broken down more quickly in the riper, squashier tomatoes because there is more pectinase in these tomatoes. But we must be careful.

5 Can you think of an explanation for your predicted results other than an increase in the amount of pectinase? If the answer is yes, you need to design a control to eliminate this possibility.

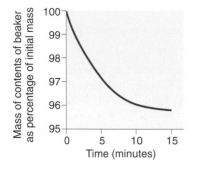

Figure 4.25

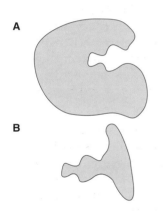

Figure 4.26

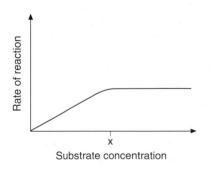

Figure 4.27

Examination questions

1 Enzymes are catalysts that catalyse specific reactions by lowering their activation energy. The lock and key and the induced fit models have been used to explain the way in which enzymes work.

(a) Explain what is meant by *activation* energy.

(1 mark)

(b) (i) Describe how the lock and key model can be used to explain how an enzyme breaks down a substrate molecule.

(3 marks)

(ii) Describe how the induced fit model differs from the lock and key model of enzyme action.

(2 marks)

Catalase is an enzyme found in many cells. It catalyses the breakdown of hydrogen peroxide:

$$\text{hydrogen peroxide} \xrightarrow{\text{catalase}} \text{water} + \text{oxygen}$$

Cylinders of potato were cut with a cork borer. The cylinders were then sliced into discs of the same thickness and put into a small beaker containing 50 cm³ of hydrogen peroxide. The mass of the beaker and its contents was recorded over a period of 15 minutes. The results are shown in Figure 4.25.

(c) Explain why the mass of the contents of the beaker fell as the reaction progressed.

(1 mark)

(d) Explain, in terms of collisions between enzyme and substrate molecules, why the rate of the reaction changed over the period of time shown on the graph.

(2 marks)

2 (a) Figure 4.26 shows an enzyme, and B is the substrate of this enzyme. By drawing on this diagram, show how a competitive inhibitor would affect the activity of the enzyme.

(2 marks)

(b) Figure 4.27 shows the effect of changing substrate concentration on the rate of an enzyme-controlled reaction.

Explain why increasing the substrate concentration above the value shown at X fails to increase the rate further.

(2 marks)

(c) Explain how adding excess substrate could overcome the effect of a competitive inhibitor.

(2 marks)

5 Gas Exchange

Throughout much of human history, people have been obsessed with climbing tall mountains. One small but important branch of biology deals with the effect of high altitude on humans, as well as other animals. All tissues and organs, but especially the brain, are affected by a lack of oxygen (Figure 5.1). Thomas Hornbein, an anaesthetist, who climbed Mount Everest, concluded that at extreme altitude or low oxygen levels 'the brain…rather than the exercising muscle is the organ ultimately limiting function.'

Extreme altitude is defined as a height greater than 5800 metres above sea level. At such heights, an organism is besieged by bitter cold, high winds, low humidity and high levels of solar and ultraviolet radiation. However, by far the most important physical challenge is hypoxia, or low levels of oxygen reaching the body tissues. The percentage of oxygen in the air actually remains the same, 21%. However, the atmospheric pressure drops steadily with increasing height, so that on top of a 3500 m mountain, for example Ben Nevis, the amount of oxygen present is only 65% that at sea level and at the summit of Mount Everest (8850 m) it is only 32% (see Figure 5.2).

At extreme altitude, most people experience a sharp decrease in appetite, breathlessness at rest, muscular fatigue, exhaustion on attempting any sort of exertion and a falling off of mental capacity. They have a significant reduction in the speed of their reactions and in their ability to make decisions. Also memory is also noticeably impaired. As a climber presses on to greater heights, brain tissue is further deprived of oxygen and bizarre mental states often occur.

Figure 5.1
To survive at high altitude humans normally require a supply of oxygen.

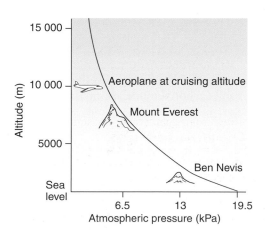

Figure 5.2
At higher altitudes, the atmospheric pressure is lower and the partial pressure of oxygen is less.

All living cells require a source of energy in order to survive. The energy is used for growth and for the processes which maintain each cell. All cells obtain their energy by the process of respiration. Respiration occurs in every cell within the body, generating adenosine triphosphate (ATP), using oxygen and producing carbon dioxide as a waste product. As a result, oxygen is depleted inside the cells whilst carbon dioxide accumulates. The diffusion of oxygen and carbon dioxide in opposite directions across the surface of a cell is called **gas exchange**.

In small animals, oxygen and carbon dioxide travel the entire distance between the external environment and mitochondria, where respiration takes place, by diffusion alone. In larger and more active animals, gas exchange over the general body surface cannot satisfy the animals' needs. One reason for this is that larger animals do not have a large enough surface area to exchange sufficient amounts of gases for the increasing demand made by a very large number of cells.

If the size of any organism increases, the relationship between its surface area and its volume changes. As Figure 5.3 shows, if you increase the size of an organism by doubling the length of each side, keeping the shape constant, the surface area (and hence the oxygen it can absorb) is increased by four times. However, its volume (and therefore the oxygen it needs) is increased by eight times. Thus by increasing body length the ability to supply oxygen by diffusion through the surface is reduced. Conversely, with a small organism, its volume decreases faster than its surface area. Thus the surface area:volume ratio is greater in small organisms.

Figure 5.3
The relationship between the size of an organism and its surface area to volume ratio.

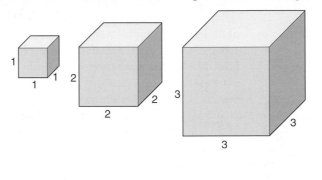

Surface area	6	24	96
Volume	1	8	64
Surface area:volume ratio	6:1	3:1	1.5:1

There are three ways in which larger animals can obtain sufficient oxygen to allow respiration to occur in all cells:

- they can have a body shape that provides a large surface area relative to their volume. Some organisms develop flat bodies, which increase their surface area

- they can develop a specialised gas exchange system, respiratory organs, which have a large surface area. These organs vary and are suited to the environment in which organisms live. They include lungs in humans and gills in fish

- they can have a blood system to transport oxygen to cells that are far from the body surface. Oxygen and carbon dioxide are carried in the blood to and from a gas exchange organ.

Structure of the lungs

As air enters the body it is filtered, moistened and warmed by different parts of the upper respiratory tract (Figure 5.4). The nostrils are lined with small hairs that act as filters to prevent large airborne objects from penetrating the lungs. The nasal cavities warm and moisten the air and collect airborne particles on a mucous layer. Tiny hair-like structures called **cilia** (see Figure 5.5) protrude from the epithelial cells that line the nasal cavities and sweep the mucous and trapped particles back towards the throat.

Figure 5.4
The human respiratory system.

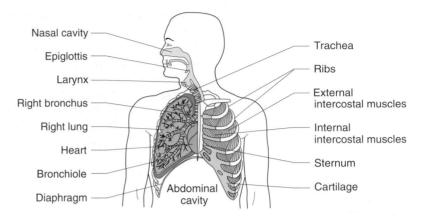

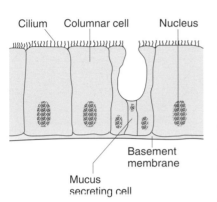

Figure 5.5 (above)
The ciliated epithelium.

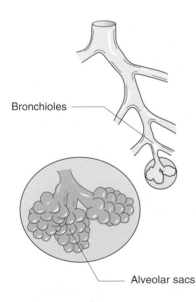

Figure 5.6 (above)
The branching network of airways leading to the alveoli.

Figure 5.7 (right)
The numerous alveoli, with a dense capillary network, where gas exchange takes place.

The nasal and buccal cavities are separated from each other; chewed food passes towards the oesophagus, while air entering the nasal cavities passes towards the trachea. Air, liquids and food pass through a chamber behind the tongue, the **pharynx**. To prevent food or liquid from entering the **trachea** during swallowing, a flap of tissue, the **epiglottis**, temporarily seals off the larynx, the box-like entrance to the trachea and lung system, so that food and liquids cannot enter the air passages.

The trachea is a muscular tube leading to the lungs. The inner surface of the trachea is lined with ciliated epithelial cells that also produce mucus, while the outer wall contains incomplete or C-shaped rings of cartilage that keep the trachea open at all times. The trachea divides into two **bronchi**, similar to trachea but with a smaller diameter. Each bronchus is supported by irregular plates of cartilage and, like the trachea, is lined with ciliated epithelia, which provide protection against microorganisms. The smallest bronchi divide still further into **bronchioles**. These contain no cartilage and are held open by the elasticity of the surrounding tissue. They have smooth muscles in their walls which enable their diameter to be controlled. These bronchial tubes form a system of hollow air ducts that end in clusters of tiny, thin-walled, blind-ending air sacs called **alveoli** (Figure 5.6). These average about 100 μm in diameter.

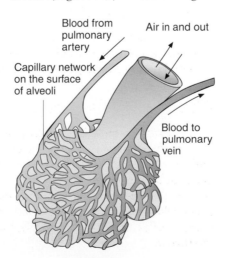

Adult human lungs contain about 750 million alveoli, with a total surface area for gas exchange of about 80 m². Within the walls of each alveolus are numerous capillaries, which receive blood from the pulmonary arteries and drain it into the pulmonary veins (Figure 5.7). So dense are the capillaries that they form an almost continuous 'pool' of blood with an area estimated to be about 87% of the alveolar surface.

Gas exchange in the lungs

The alveoli are the sites of gas exchange and may be regarded as the functional unit of the lungs. The walls of the alveoli are extremely thin, usually only one cell thick, and consist of squamous epithelia. Each alveolus is surrounded by a dense bed of blood capillaries (Figure 5.8).

Figure 5.8
Photomicrograph of a section through several alveoli and capillaries.

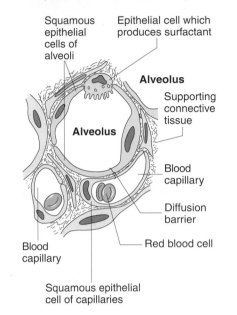

Table 5.1 shows the amount of oxygen and carbon dioxide in the air moving in and out of the lungs. Other gases, such as nitrogen and water vapour, are also present. The percentage of nitrogen hardly varies, but the amount of water vapour in exhaled air is usually greater than that in inhaled air.

	Oxygen	Carbon dioxide
Inspired air	20	0.04
Alveolar air	13	5
Expired air	15	4

Table 5.1 The percentage of oxygen and carbon dioxide in inspired air, alveolar air and expired air.

Oxygen entering the alveolus dissolves in the film of water on its wall and then moves by diffusion across cells to the blood. As you can see from Figure 5.9, the blood flowing to the alveolus has a higher concentration of carbon dioxide and a lower concentration of oxygen than alveolar air. Under these circumstances oxygen and carbon dioxide diffuse down their respective concentration gradients: oxygen from the alveoli diffuses into the blood and carbon dioxide from the blood diffuses into the alveoli.

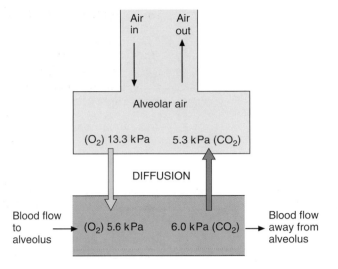

Figure 5.9
The exchange of oxygen and carbon dioxide that takes place as blood flows past an alveolus.

The movements of these gases are cases of simple diffusion and require no energy. There are actually few instances where the process of simple diffusion takes place within the body, but this is the case in the lungs. As a result, when the partial pressure of oxygen in the atmosphere falls below normal, oxygen deprivation rapidly develops. This is because the reduced concentration gradient slows the rate of diffusion and passive transport is insufficient to meet oxygen demands.

Once in the blood plasma, most oxygen passes into red blood cells, where it binds to haemoglobin and is transported through the body.

An important feature of multi-cellular animals is the presence of a gas exchange organ with a large enough surface area for sufficient intake of oxygen and release of carbon dioxide to occur to meet the demands of all body cells. However, as the body's requirement for energy and therefore oxygen increases, the rate of gas exchange must also increase. The large gas exchange organ with resulting large moist surface area has its limitations because the larger the surface area, the more water the body loses by evaporation. A compromise must be reached so that the gas exchange surface remains moist and permeable, and so these surfaces tend to be internal. Larger organisms have developed an impermeable outer layer and do not use their body surface for gas exchange.

Q 1 Suggest why the gas exchange structures in air breathing animals tend to be internal?

2 Suggest why gas exchange surfaces are moist?

Fick's law

Fick's law of diffusion states that the rate of diffusion of a gas is directly proportional to the area of the diffusion surface and to the difference in its concentration, but is inversely proportional to the thickness of the exchange surface or the distance over which it occurs:

$$\text{diffusion rate} \propto \frac{\text{surface area} \times \text{difference in concentration}}{\text{thickness of surface}}$$

This law provides an effective framework for considering how the maximum rate of diffusion of oxygen and carbon dioxide in the lungs is achieved.

To obtain the maximum rate of diffusion, a gas exchange system must have the following features:

- a large surface area
- the smallest possible diffusion pathway
- a large difference in concentration.

In a human lung these are achieved in the following way.

A large surface area

It may seem puzzling that the large surface area of the lung is achieved by having several million small alveoli rather than simply increasing the size of a few. Since the alveolar cell surface is the site of gas exchange between the air and the blood, a larger relative surface area allows a more rapid rate of diffusion of oxygen and carbon dioxide.

As explained earlier, when the length of each side of a cube doubles, the volume increases more rapidly than does the surface area. Hence the surface area:volume ratio goes down.

However, if a large cube $(4 \times 4 \times 4)$ is cut into 64 small cubes $(1 \times 1 \times 1)$ the total surface area becomes four times as large (see Figure 5.10). If these were alveoli, the 64 small air sacs would be better able to diffuse oxygen and carbon dioxide between the blood and the air in the lung than one large air sac.

Figure 5.10
The effect on surface area and surface area to volume ratio of dividing a large cube into smaller ones

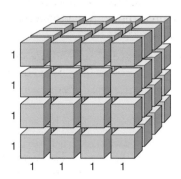

	One large cube	64 small cubes
Volume	64	64
Surface area	96	384
Surface area: volume ratio	1.5:1	6:1

A small diffusion pathway

This is caused by the presence of a special type of cell, **squamous epithelium,** found in both the alveolar wall and the capillary. These cells are extremely thin. A small diffusion pathway is achieved because both the alveolar and capillary cells are so thin that the distance between the air in the lung and the blood is very short (Figure 5.11). In fact the average distance the gases travel is only about 0.5 μm

A large difference in concentration

This is maintained in two ways. Firstly, the partial pressure of oxygen in the alveoli is always greater than that of oxygen in the capillaries because of breathing, whereby a fresh supply of oxygen is brought to the alveoli. Secondly, the blood in the capillary is always moving, removing oxygenated blood and replacing it with deoxygenated blood. Thus it can be seen that the fine structure of the lungs is superbly adapted for gas exchange.

Q 3 What would happen to the oxygen concentration gradient across the gas exchange surface separating your lungs from your blood if the air in your lungs were not replaced frequently?

Figure 5.11
The diffusion pathway. Oxygen diffuses from the alveolus into the capillary. Carbon dioxide diffuses from the capillary to the alveolus.

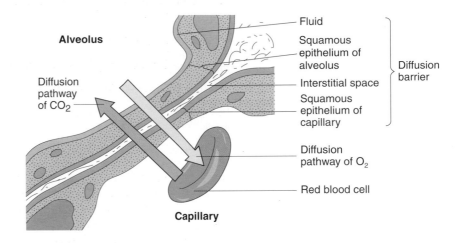

Front view of rib cage and diaphragm

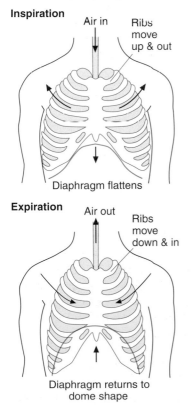

Figure 5.12
Diagram showing the movement of the ribs and diaphragm during inspiration and expiration.

The breathing mechanism

Humans fill and empty their lungs by means of a group of bones and muscles that work together. The human lungs lie within the **thoracic cavity**, a space enclosed by the ribs and separated from the abdominal cavity, which lies below by the **diaphragm**. In a human, the ribs curve forward from the backbone and meet at the breastbone. A diagonal set of muscles, the **intercostal muscles**, stretch between the ribs.

Breathing involves the alternate increase and decrease of air pressure in the lungs relative to that outside. A fall in air pressure in the lungs causes **inspiration** or breathing in; a rise in pressure in the lungs causes **expiration**, or breathing out.

Inspiration

Both the **external intercostal muscles** and the diaphragm muscles contract during inhalation of air (Figure 5.12). As the external intercostal muscles shorten, the rib cage moves up and out. At the same time, the diaphragm, which is dome-shaped at rest, flattens and pulls downward.

During inspiration, these muscular actions expand the volume of the thoracic cavity, lowering the air pressure inside the cavity below atmospheric pressure. As a result, air flows in through the nostrils, down the trachea, bronchi and bronchioles and into the alveoli.

Expiration

Expiration is largely a passive process brought about by relaxation of the external intercostal muscles and the diaphragm. The energy for expiration thus comes from energy stored as elastic tension generated during the previous inspiration. This relaxation is called elastic recoil. It results in the rib cage moving down and in, and the diaphragm returning to its original dome shape. These movements decrease the volume of the thoracic cavity, increasing the air pressure inside the cavity above atmospheric pressure and as a result air flows out.

During active exercise, however, a second set of muscles between the ribs, the **internal intercostal muscles**, contract and forcibly lower the rib cage. This expels more air from the lungs, making room for a larger volume of air with each inhalation.

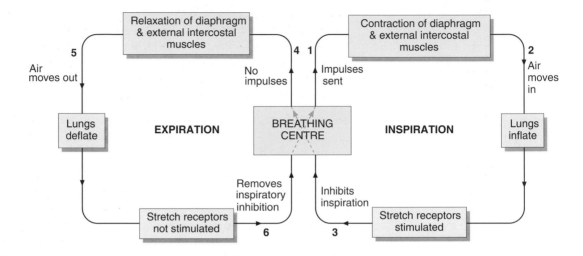

Figure 5.13
The coordination of breathing.

Coordination of breathing

Although we can breathe in or out voluntarily, breathing is largely automatic. The control centre for these automatic movements lies in the **breathing centre,** which is in the portion of the brain called the medulla. It is divided into two regions: an **inspiratory centre** and an **expiratory centre** (Figure 5.13).

When the inspiratory centre sends impulses to the external intercostal muscles via the intercostal nerves and to the diaphragm via the phrenic nerves, these muscles contract and cause inhalation.

As the lungs expand, stretch receptors in the walls of the bronchial tree are stimulated. These send impulses via the vagus nerve to the expiratory centre, which automatically cuts off inspiratory activity. This results in the diaphragm and external intercostal muscles relaxing, causing expiration. When the muscles have relaxed, the stretch receptors in the lungs are no longer stimulated therefore the expiratory centre becomes inactive and inspiration begins again. The inspiratory centre can once more send impulses to the external intercostal muscles and the diaphragm, and inhalation occurs again. This cycle is repeated again and again.

Control of breathing

It is well known that both the rate and depth of breathing increase during exercise and decrease during sleep. The ventilation rate must therefore match the oxygen needs of the body. The breathing centre must have information concerning the activity of the body, but what is being monitored and where are the receptors?

Q 4 Explain why it is not possible to kill yourself by holding your breath?

5 Explain how hyperventilation is able to suppress the urge to breathe.

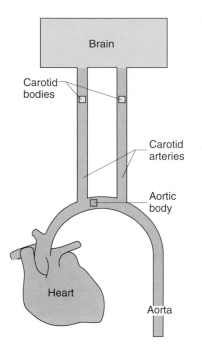

Figure 5.14
The aortic and carotid bodies.

There are two possible stimuli:

- carbon dioxide concentration: levels in the blood go up when the rate of respiration increases and more carbon dioxide is produced as a waste product

- oxygen concentration: levels in the blood go down as it is used in respiration to produce extra ATP as an energy source for exercise.

It has been shown experimentally that the oxygen level in the air has to fall from 20% to 12% before an increase in breathing rate is stimulated. In contrast, a tiny increase in the proportion of carbon dioxide in inhaled air strongly stimulates an increase in breathing. It appears therefore that the body is much more sensitive to changes in carbon dioxide concentration than to changes in oxygen levels.

In air, oxygen is normally abundant and therefore it is not likely to be a limiting factor, however, carbon dioxide removal is a problem. Thus, in humans carbon dioxide level in the body mainly drives the breathing centre.

The most important site for carbon dioxide detection is the breathing centre itself, situated in the medulla of the brain. A small group of cells, **chemoreceptors**, monitor the carbon dioxide and pH levels of the fluid within the brain. Two further small clusters of chemoreceptors, located in the arch of the aorta and in the wall of the carotid arteries, are known as the **aortic and carotid bodies** (Figure 5.14). These are stimulated by a rise in the level of carbon dioxide, a fall in pH and a fall in the level of oxygen in the blood. The breathing centre receives information as a nerve impulse from these chemoreceptors and uses this to regulate breathing rate.

In response to an increase in the concentration of carbon dioxide or a fall in pH (caused by carbon dioxide dissolving in the blood plasma, producing a weak acid) the inspiratory centre increases the rate of nerve impulses sent to the inspiratory muscles. This results in an increase in both the rate and depth of breathing.

The regulation of breathing is an example of **homeostasis**. When the concentration of carbon dioxide rises, this stimulates an increase in breathing rate, which in turn reduces the concentration of carbon dioxide, leading to a decrease in breathing rate (Figure 5.15).

Figure 5.15
The homeostatic control of carbon dioxide.

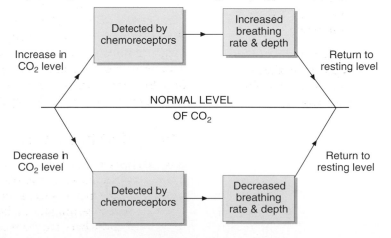

Ventilation cycle

Q 6 If the average breathing rate is 15 breaths per minute and the tidal volume is 0.5 litres. Calculate the ventilation rate.

A person at rest who is breathing normally takes in and expels about half a litre of air during each ventilation cycle. This is known as the **tidal volume** and it can be measured and recorded using a spirometer.

The rate at which a person breathes is called the **ventilation rate**. This is usually expressed as the volume of air breathed per minute:

ventilation rate = tidal volume × number of breaths per minute

The ventilation rate changes according to the circumstances: in muscular exercise, for example, both the frequency and depth of breathing increase, resulting in a higher ventilation rate. Lungs have a much greater potential volume than is ever used in resting conditions and this allows the ventilation rate to adapt to changing needs. If you take a deep breath, you can take into your lungs an extra 3 litres of air over and above the tidal volume. This is called the **inspiratory reserve volume** and is brought into use when required. If at the end of a normal expiration you expel as much air as you possibly can, the extra air expired is approximately 1 litre and is called the **expiratory reserve volume**. The total volume of air that can be expired after maximum inspiration is known as the **vital capacity**. The vital capacity of an average person is between 4 and 5 litres but that of a fit athlete may exceed 6 litres. Even after maximum expiration, approximately 1.5 litres of air still remain in the lungs. This is known as the **residual volume**. The various lung volumes are shown in Figure 5.16.

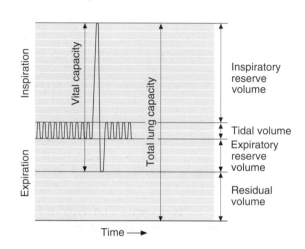

Figure 5.16
The typical lung volumes of humans.

Of the half litre inspired in quiet breathing, only about 350 cm^3 gets into the parts of the lung where gas exchange takes place. The rest remains in the trachea and bronchial tubes, collectively known as the 'dead space', where no gas exchange takes place.

Extension box 1

Consider how cigarette smoke affects the lungs. Chemicals in the smoke from just one cigarette immobilise cilia in the bronchioles for several hours. The particles in smoke also stimulate mucus secretions that in time block the airways. Other chemicals in smoke can kill protective white blood cells present in the respiratory tract. What starts as 'smoker's cough' can end in bronchitis and emphysema. One in five of all smokers develop the crippling disease emphysema, which cannot be cured as the damage caused to the lungs cannot be reversed.

Obstructive lung disease, which includes bronchitis and emphysema, is the name given to a number of lung conditions in which the passage of air is obstructed. These two conditions frequently exist together, bronchitis being a disease of the bronchi and emphysema a disease of the bronchioles and alveoli. In emphysema the alveoli become permanently and abnormally enlarged due to destruction of alveolar walls.

Figure 5.17
The balance between elastase and the elastase inhibitor in a normal lung and a lung with emphysema.

Healthy lungs contain large amounts of elastic connective tissue, mainly a protein called **elastin**, which makes the lung tissue elastic so that it can expand and contract as we breathe in and out.

The development of emphysema appears to be related to the amount of the enzyme **elastase** that acts on lung tissue. The greater the amount of elastase, the more elastin is broken down, and large holes appear in the lungs.

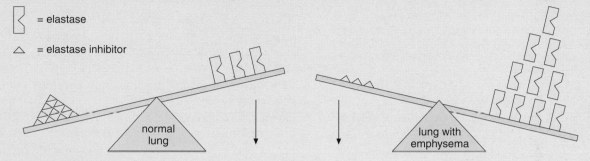

Within our lungs we have a protein called a **proteinase inhibitor** (PI) and its main action is to inhibit the enzyme elastase. In healthy lungs elastin is not broken down because the PI inhibits the action of the enzyme.

In emphysema, lung tissue is destroyed because the normal balance between elastase activity and anti-elastase activity is shifted in favour of elastase (Figure 5.17).

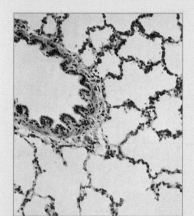

In lungs with emphysema, the elastin, which can be compared to elastic bands, has perished and become permanently stretched. The lungs are no longer able to force out all the air from the alveoli. The residual air left in the alveoli prevents fresh air, taken in with each breath, from reaching the alveolar walls and passing through to the blood.

The stretched and damaged alveoli are often referred to as 'dead space' because little if any exchange of gas can take place across them. As the enzyme elastase can degrade many other proteins besides elastin, there is breakdown and loss of other lung tissue too until, in extreme cases of the disease, the lung is composed of rather large holes that are non-functional (Figure 5.18).

The loss of alveoli has two important consequences. There is a decrease in surface area, resulting in reduced gas exchange, and with the loss of the elastic fibres in the alveolar walls, the lungs have a decreased ability to recoil and expel air.

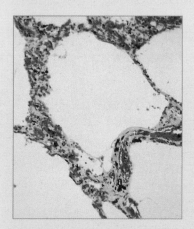

The only way to minimise the chance you have of developing emphysema is not to smoke at all or give up; lung function cannot be restored to smoke-damaged lungs but giving up smoking does significantly reduce the rate of further deterioration of lung function.

Figure 5.18
Photograph of a normal lung (top) and a lung with emphysema (bottom).

Extension box 2

We breathe air, which is a mixture of gases. Atmospheric air at sea level contains approximately 20% oxygen and 80% nitrogen by volume and exerts a total pressure of about 100 kPa. Of this 100 kPa only 20 kPa is due to oxygen molecules; the remaining 80 kPa is due to nitrogen. These are the partial pressures of the individual gases and are symbolised as pO_2 and pN_2 (Figure 5.19).

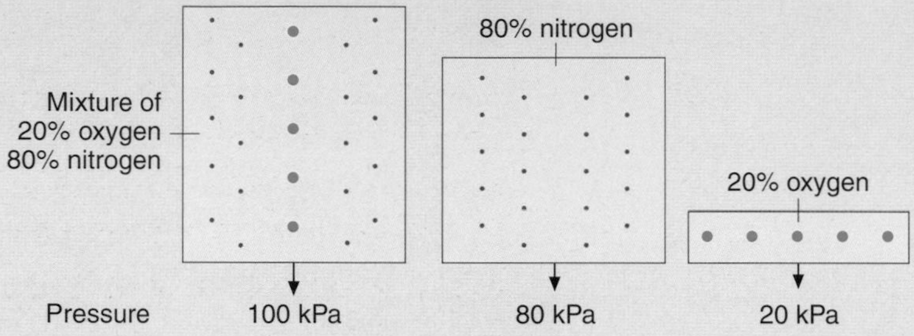

Figure 5.19
The partial pressures of oxygen and nitrogen in atmospheric air.

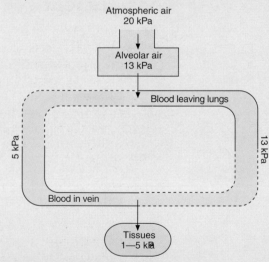

Mixture of 20% oxygen 80% nitrogen

80% nitrogen

20% oxygen

Pressure   100 kPa   80 kPa   20 kPa

Partial pressure is the driving force causing the transfer of oxygen from air to blood in the alveoli of the lungs and from blood to mitochondria in the tissues of the body.

The partial pressure falls at each stage of transfer: from atmosphere to lung air, to blood, to tissue fluid, to cytoplasm to mitochondria (Figure 5.20).

Atmospheric air
20 kPa

Alveolar air
13 kPa

Blood leaving lungs

5 kPa

13 kPa

Blood in vein

Tissues
1—5 kPa

Figure 5.20
The fall in partial pressures of oxygen during the transfer of oxygen from air to blood.

Why partial pressure and not just percentages?

This is because oxygen uptake is affected by atmospheric pressure. Even at sea level, atmospheric pressure is not always exactly 100 kPa and at the top of Mount Everest the pressure can vary greatly from this. Here the air still contains 20% oxygen by volume but due to a very low atmospheric pressure, the partial pressure of oxygen entering the lungs is only 5.8 kPa compared with 20 kPa at sea level. Thus the concentration gradient of oxygen between the lungs and the blood is greatly reduced.

Summary

- The human gas-exchange system consists of the nasal cavity, trachea and lungs. Movement of the lungs is effected by the intercostal muscles.

- Gas exchange takes place across the alveolar epithelium.

- The alveoli provides a large surface area for gas exchange.

- In the alveoli, oxygen diffuses from the alveolar air to the blood, and CO_2 diffuses from the blood into the alveolar air.

- Fick's law provides a framework to show how the maximum rate of diffusion of respiratory gases is achieved.

- The medulla and the phrenic nerves generate a basic breathing rhythm

- Pulmonary ventilation is the product of tidal volume and breathing rate.

- Exercise increases pulmonary ventilation.

Assignment

Biologists are like other scientists. They need to be able to carry out simple calculations in order to summarise and analyse data, and interpret the results. In this assignment we shall look at the relationship between size and surface area, and the way in which it affects the diffusion of respiratory gases and the transfer of heat. In carrying out these exercises you will have the opportunity to practise some of the skills necessary for your Key Skills portfolio concerning the application of number. You will be required to:

- carry out multistage calculations to do with:

    - amounts and sizes
    - scales and proportions
    - handling statistics
    - rearranging and using formulae

- interpret the results of calculations, present findings and justify methods.

It is not easy to find the surface area of an animal such as a rabbit or a cow. Various attempts have been made by measuring skins that have been removed from the animals concerned or by comparing them with a series of cylinders and cones. Perhaps the most imaginative approach involved using a roller to paint a cow and calculating the surface area from the number of revolutions of the roller!

1 Explain why there are likely to be errors in finding an animal's surface area by:

(a) removing its skin and measuring this

(b) comparing the animal with a series of cylinders and cones.

The simplest way to investigate the relationship between size and surface area is not to consider an animal but to look instead at a simple geometrical shape such as a cube.

2 (a) Copy and complete the table below. This will allow you to compare a number of cubes which have sides of different lengths. For each cube, you will need to calculate the total surface area (remember, a cube has six faces), volume and the ratio of surface area to volume. For simplicity, units have been omitted.

Length of one side	Total surface area	Volume	Ratio of surface area to volume
1			
2			
3			
4			
5			
6			

Table 5.2

(b) Which of the following statements are true and which are not true about the information in the table:

A The larger the cube, the greater its surface area.
B Small cubes have small surface areas in relation to their volumes.
C As cubes get larger, their volumes increase faster than their surface areas.
D An increase in the size of a cube is associated with a decrease in the ratio of surface area to volume.

The graph in Figure 5.21 shows measurements of the body mass and surface area of a large number of different species of vertebrate animals.

Figure 5.21
Graph showing the relationship between surface area and body mass in different species of vertebrate.

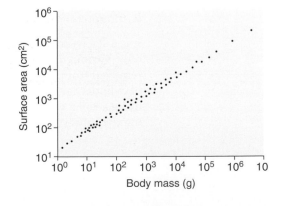

3 (a) A log scale was used to plot the data in Figure 5.21. This means that each point marked on the axes is 10 times larger than the previous point. What is the advantage of using a log scale when plotting these data?

(b) Use the graph to estimate:
   (i) the surface area of an animal that weighs 1 kg
   (ii) the body mass of an animal that has a surface area of 10 m².

The relationship between surface area and the size of an animal, measured in terms of either its volume or its mass, is important when considering the transfer of heat. The two animals shown in the photographs in Figures 5.22 and 5.23 both live in tropical regions. Use the information in your table and that in Figure 5.21 to answer the questions under each photograph.

Figure 5.22

4 What is the advantage of large ears to an African elephant?

5 Large reptiles such as the crocodile (see page 88) are found in the tropics. Use your knowledge of heat transfer and body size to suggest why large reptiles could not live in colder regions.

Figure 5.23

The relationship between size and surface area is also important when considering the diffusion of substances into and out of cells. Figure 5.24 shows a photograph of a cell from the lining of the small intestine.

Figure 5.24

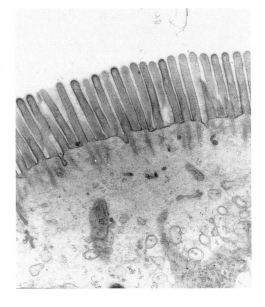

6 The relationship between real size in a photograph and the magnification is given by the formula:

$$\text{real size} = \frac{\text{size in photograph}}{\text{magnification}}$$

Use this formula to calculate the magnification of the photograph. Show your working, taking the length of one microvillus to be 3μm.

7 This cell absorbs molecules from digested food. By how many times do the microvilli increase the surface area of the cell that is in contact with digested food?

Note:

● There are no instructions to tell you how to do this. However you go about this you should remember that all you can really do is to make a good estimate. Because of this you are not justified in calculating your answer to several decimal places.

● Set out your calculations as clearly as you can, giving your reasons for carrying out each step. Describe any assumptions you need to make.

Examination questions

1 Figure 5.25 shows a section through some alveoli in the lung of a human.

Give **three** features visible in the figure that help to increase the rate of diffusion across the wall of the alveoli into the blood.

(3 marks)

2 Fick's law can be given as:

diffusion rate is proportional to $\dfrac{\text{surface area} \times \text{difference in concentration}}{\text{thickness of surface}}$

(a) With reference to gas exchange in the human lung:

 (i) describe two features that ensure a large surface area for gas exchange.

(2 marks)

 (ii) give two processes that ensure that a difference in concentration is maintained.

(2 marks)

(b) Figure 5.26 shows the use of the mouth-to-mouth method of resuscitation (ventilation). Suggest how it is possible for exhaled air to be effective when this method is used.

(2 marks)

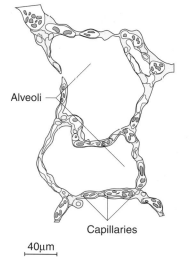

Alveoli

Capillaries

40μm

Figure 5.25

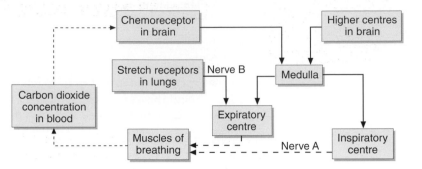

Figure 5.26 (above)
Figure 5.27 (right)

3 Figure 5.27 shows some parts of the mechanism that controls breathing.

(a) Which muscles are stimulated by Nerve A?

(2 marks)

(b) Explain the role of Nerve B in the control of breathing when a subject is at rest.

(1 mark)

(c) Give one example of the role of higher centres of the brain in the control of breathing.

(1 mark)

(d) Table 5.3 shows the effect of variations in breathing rate and tidal volume on alveolar ventilation (fresh air reaching the alveoli per minute). The dead space of the respiratory system, i.e. the volume of the trachea and bronchi through which no respiratory exchange can take place, is 150 cm^3.

	Column X	Column Y
Breathing rate (breaths per minute)	30	10
Tidal volume (cm^3)	200	600
Pulmonary ventilation rate (cm^3 per minute)	6000	6000
Alveolar ventilation rate (cm^3 per minute)	$(200 - 150) \times 30 = 1500$	$(600 - 150) \times 10 = 4500$

Table 5.3

(i) Describe how a person would be breathing to produce the figures in column X.

(1 mark)

(ii) During exercise, both the breathing rate and tidal volume increase. Use the information above to explain why an increase in the breathing rate alone would not be able to satisfy the demands of the body during exercise.

(3 marks)

The Heart and Circulation

The human heart is a remarkable organ. It beats over 100 000 times a day, that's around two and a half thousand million times in a lifetime and it never stops for a rest. To do this, the heart muscle must maintain a high rate of respiration and receive sufficient oxygen and glucose.

A heart attack occurs when one of the coronary arteries becomes blocked. Part of the heart muscle is starved of oxygen and dies. Sometimes a heart attack affects the pacemaker and other parts of the system by which the heartbeat is controlled. This results in heart block: a condition which, if not treated, is fatal. An artificial pacemaker may be fitted. This consists of a thin wire that is threaded along one of the veins in the neck or arm into the right side of the heart where its tip is embedded in the wall of the ventricle. The other end of this wire is attached to an artificial pacemaker, which is adjusted so that the ventricle beats at between 60 and 80 beats per minute.

There are some much more advanced pacemakers. They measure changes in breathing rate, body temperature and vibration, and alter the rate at which the artificial pacemaker is causing the heart to beat. When the person takes some exercise, the pacemaker makes sure that the heart beats at an appropriate rate.

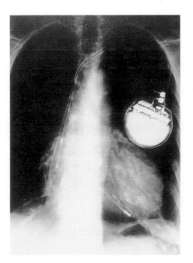

Figure 6.1
This person has been fitted with an artificial pacemaker. You can see the pacemaker and the wires leading to the heart. Apart from regular visits to the hospital to check that the pacemaker is still working, the patient can expect to live a normal life.

key term

The **coronary arteries** supply the heart muscle with oxygen and glucose.

In Chapter 2, we looked at ways in which different substances move in and out of cells. One of these was diffusion. Diffusion is very efficient at transporting substances over short distances but it is nowhere near as effective when greater distances are involved. In a mammal, there are many surfaces where substances are exchanged between organs and their surroundings: respiratory gases are exchanged in the alveoli of the lungs, nutrients obtained from the digestion of food are absorbed by the cells lining the gut, and muscle cells exchange respiratory gases, obtain their glucose and get rid of various waste products. These processes take place over very short distances and frequently involve diffusion. However, we need a system that will link the different exchange surfaces and allow the rapid bulk movement of substances around the body. This is the role of the blood. Because of what it does, it is sometimes described as a **mass transport** system, transporting large quantities of different substances from a **source** to a **sink**.

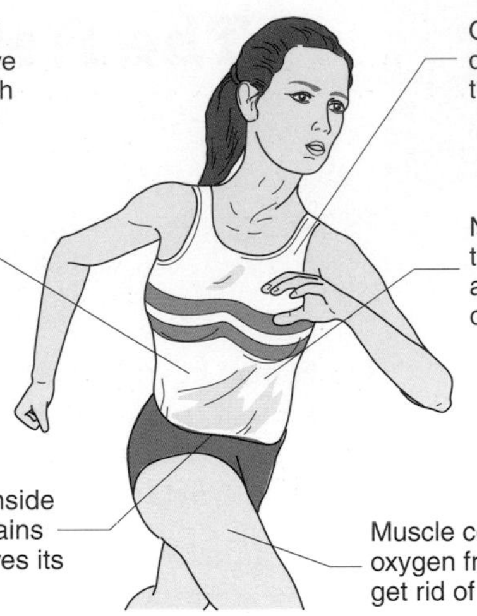

Figure 6.2
The many exchange surfaces in
the body of a human are linked
by the blood system, which acts
as a mass transport system.

One of the functions of the kidneys is to remove excretory products such as urea from the blood

Oxygen and carbon dioxide are exchanged in the alveoli of the lungs

Nutrients obtained from the digestion of food are absorbed through the wall of the small intestine

The fetus developing inside its mother's uterus obtains its nutrients and removes its waste via the placenta

Muscle cells obtained glucose and oxygen from the blood. They also get rid of their waste products

Q 1 Oxygen and glucose are transported around the body in the blood system. For each of these two substances, name one organ that acts as a source and one that acts as a sink.

In this chapter, we shall look at the role of blood cells and different sorts of blood vessels in transporting substances between exchange surfaces. We shall also consider the ways in which the structure of the heart is related to its function, knowledge of which has helped us to develop new ways of treating heart disease, such as the use of artificial pacemakers. Finally, we shall see that the blood system is much more than a series of pipes running between different organs. Flow through the different vessels is constantly being adjusted to meet the needs of the body.

The circulation of blood

A mammal's heart serves as a pump. It pumps blood out through arteries. These branch into smaller arterioles and then into capillaries. Capillaries are tiny vessels with very thin walls that allow the exchange of substances between the blood and the cells of the body. Blood is collected from the capillaries by a network of venules. It flows from these venules into the larger veins and back to the heart.

The larger blood vessels in the body share the same basic structure. Their walls have three layers:

- a thin inner lining layer, or endothelium, that is very smooth and enables blood to flow through vessels without causing friction
- a middle layer containing muscle and elastic fibres
- a tough outer layer that helps to protect the vessel from the pressure exerted by other organs rubbing against it.

Table 6.1 shows how the relative thickness of these layers varies and depends on the type of blood vessel involved. The blood contained in the larger arteries has just left the heart. It is at a high pressure and there is a surge every time the ventricles of the heart contract. As the blood is forced along the artery, the elastic fibres in the middle layer are stretched, allowing the wall to bulge outwards. Between the surges in pressure, these elastic fibres recoil. This helps to force the blood along the artery and gradually evens out the flow of blood.

Vessel	Appearance	Inner layer	Middle layer	Outer layer
Artery		Present	Present as a thick layer containing muscle and many elastic fibres	Present
Capillary		Present	Absent	Absent
Vein		Present	Present but thinner than in an artery. Contains muscle and a few elastic fibres	Present

Table 6.1

Q 2 Explain why you can feel a pulse if you place your fingers over a large artery in the wrist or neck but not if you place them over a vein.

The amount of blood required by different parts of the body varies. If you take some form of fairly vigorous exercise, the blood supply to your muscles and skin will increase, while that to your digestive system will decrease. The presence of muscle in the middle layer of the walls of the smaller arteries and arterioles allows the diameter of these blood vessels to change and the blood supply to various organs to be constantly adjusted to meet the needs of the body.

In the capillaries, substances are exchanged between the blood and the surrounding cells. Capillary walls consist only of the thin endothelium and, as a result, they are permeable, that is water and many substances with small molecules are able to pass through them. Some of the white cells in the blood can squeeze between the endothelial cells and leave the blood, particularly at sites of infection.

Once the blood has drained back into the veins it is at a relatively low pressure. Look at the giraffe in Figure 6.3. There are a number of aspects of the blood system which enable the blood to be returned to the heart. These include:

Figure 6.3
A number of different features allow blood to be returned from the feet of a giraffe to its heart.

- valves in the veins (see Figure 6.4) that prevent backflow

- muscles that surround the veins. As an animal such as this giraffe walks, these muscles contract and squash the veins. This squeezes the blood along them. The presence of the valves ensures that blood can only be squeezed in one direction – towards the heart

- breathing in creates a negative pressure in the chest. This helps to draw blood into the heart from the veins.

Figure 6.4
The valves in veins are rather like the hip pocket in a pair of jeans. If you run your hand upwards, it squashes the pocket flat and your hand moves smoothly upwards. If you run your hand down, it pushes the pocket open and stops your hand moving further.

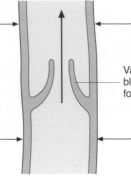

When the muscles here contract the increased pressure of the blood causes the valves to shut. This prevents the backflow of blood

Valve open allowing blood to flow forwards heart

When muscles here contract, the veins are squashed. This squeezes the blood along

Q 3 A person has an accident and cuts through a major blood vessel in the wrist. Explain how you could tell if the escaping blood was coming from an artery or from a vein.

Extension box 1

Some mammals live in extremely inhospitable conditions. In places such as the deserts of Arabia and Africa, daytime temperatures can rise as high as 45°C. Yet these regions are not devoid of life. Far from it. There are many mammals found in such regions. Some are small and can either rest in the shade or remain underground during the heat of the day. Larger animals such as antelopes find this solution impossible. They have complex physiological mechanisms that allow them to tolerate high temperatures.

Suppose we look at what happens when an antelope such as an oryx (see Figure 6.5) is exposed to very hot and dry conditions. It doesn't sweat as we do. Instead of staying constant, its body temperature rises. This would be fatal in a human. The advantage to the antelope of this behaviour is that valuable water is not lost as sweat in trying to keep a constant body temperature. Not all organs in the body, however, can tolerate such high temperatures. The animal cannot survive if the temperature of the brain is allowed to rise in this way. So how is the blood supply to the brain cooled? This is where the special pattern of circulation in the nose comes in.

An antelope exposed to these very hot conditions does not sweat but it pants. This results in moisture evaporating from the membranes in the nose. Heat energy is transferred from the body in turning the water on these surfaces into vapour. The inside of the nose will therefore be cooler than the rest of the body. This means that the blood flowing back through the veins may be several degrees cooler than that in the rest of the body.

At the base of the brain there is a branching network of blood vessels. In this network, the warm arterial blood flows past the cooler venous blood. Heat is lost from the blood in the arteries to that in the veins. This serves to cool the blood supply to the brain. The blood flow in the two vessels is in opposite directions and this makes the transfer of heat more efficient. It has been shown that the blood supplying the brain may be cooled by as much as 3°C in this way.

Figure 6.5
Oryx are antelope found in the semi-desert conditions of Africa and Arabia. When oryx were exposed to daytime temperatures of 40°C and night-time temperatures of 22°C, their body temperatures showed a 7°C fluctuation.

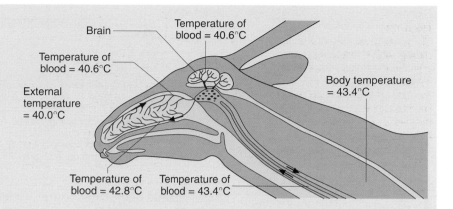

Figure 6.6
The arrangement of the blood vessels ensures that the brain of an antelope is kept cool in very hot conditions.

So, do mammals which live in deserts have big noses? This certainly ought to be the case, as the larger the nose, the greater will be the evaporative surface that cools the venous blood. It seems to be true. The surface area of the nasal sinuses of breeds of cattle and sheep that live in hot dry conditions has been found to be much greater than those of animals living in less dry conditions.

Capillaries and exchange surfaces

There are over 80 000 km of blood vessels in the body of an average human adult, and most of them are capillaries. In order to function efficiently, the individual cells in the organs that make up the body must be supplied with the oxygen, glucose, amino acids and other substances they need, and waste products such as carbon dioxide must be removed. Within almost every organ is a network of capillaries that supply or remove these substances (Figure 6.7).

The cells in the body are surrounded by **tissue fluid**. This tissue fluid is continually formed from and reabsorbed into the blood in the capillary network. Figure 6.8 shows part of a single capillary. Blood flowing into this capillary from an arteriole has a high **hydrostatic pressure**. This is the pressure brought about by the pumping action of the blood and it tends to force water and substances with small molecules out through the permeable walls of the capillary. At the same time the blood contains soluble globular proteins, called **plasma proteins**, and other dissolved substances. These lower the water potential of the blood and tend to cause water to be drawn back into the capillary by osmosis. At the arteriole end of the capillary, the effect of the hydrostatic pressure is greater than that of the water potential so fluid tends to be forced out. This fluid is very similar in composition to blood plasma but it lacks the plasma proteins, which are too large to escape.

The path that blood takes through a capillary network is controlled by small rings of muscle called **sphincters**. Opening and closing these sphincters enables blood flow to meet the needs of the organ

Arteriole brings blood from the heart

Shunt vessel joins arteriole directly to venule. Allows blood to bypass capillary network

Venule returns blood to the heart

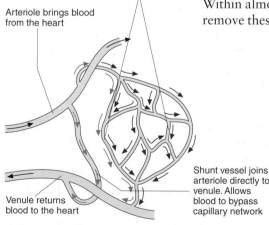

Figure 6.7
The capillary network is like a road system in a city. If you want to travel from one side of a city to the other, you can bypass the city on a motorway (the green arrows through the shunt vessel), you can use the ring road (the blue arrows through the capillaries) or you can go through the side streets (the red arrows through the smaller capillaries).

Figure 6.8
The formation and reabsorption of tissue fluid. At the arteriole end, the high hydrostatic pressure results in fluid being forced out of the capillary. At the venule end water is reabsorbed as a result of the water potential of the blood plasma.

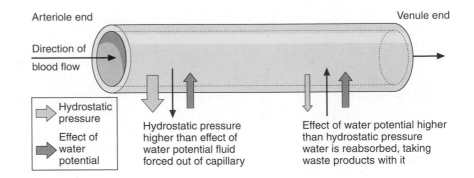

Arteriole end

Venule end

Direction of blood flow

Hydrostatic pressure

Effect of water potential

Hydrostatic pressure higher than effect of water potential fluid forced out of capillary

Effect of water potential higher than hydrostatic pressure water is reabsorbed, taking waste products with it

At the far end of the capillary, the venule end, the hydrostatic pressure is lower. This is mainly the result of some of the fluid being forced out of the blood. As the volume falls, so does the hydrostatic pressure. However, the water potential has changed very little because the plasma proteins are too large to get through the capillary walls. The effect of the water potential is now greater than that of the hydrostatic pressure and water tends to be drawn back into the blood, taking with it waste products produced by the cells.

Q **4 Explain how the following may affect the formation and reabsorption of tissue fluid:**
   (a) high blood pressure
   (b) a fall in the amount of plasma protein as a result of prolonged illness.

Over the course of a day, more fluid is forced out of the capillaries than is reabsorbed back into them. This accumulated tissue fluid is returned to the blood by the lymphatic system. Excess tissue fluid drains into small, blind-ending tubes called **lymphatic capillaries**. These lead into larger **lymph vessels**, which finally empty the lymph that they contain into the blood in the veins in the neck. Lymph vessels possess valves that ensure that their contents only flow in one direction.

Extension box 2

Sometimes more tissue fluid is formed than can be reabsorbed or removed by the lymphatic system. It accumulates and the tissues swell. This is known as **oedema**. There are many factors that may cause oedema. Some of them, like the slight swelling that takes place in the fingers in very hot weather, are of no concern in an otherwise healthy person. However, oedema may suggest that there is something much more seriously wrong with the body.

Figure 6.9
A child suffering from kwashiorkor. Note the typical swelling. This is oedema.

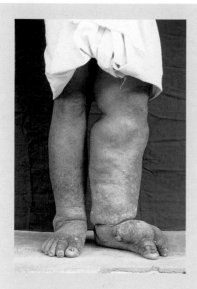

Figure 6.10
This person is suffering from filariasis. Parasitic worms have blocked the lymph vessels in his leg. Tissue fluid has accumulated and has produced permanent oedema. Over a period of time other changes occur, including the thickening of the skin which gives the condition its common name of elephantiasis – the affected leg looks like that of an elephant.

In many parts of the world malnutrition is all too common. The symptoms of **kwashiorkor** often become apparent in the second year of a child's life when he or she is weaned on to a diet consisting mainly of carbohydrate (cassava, plantain or yam) mashed up with water. This diet contains very little protein. One of the effects of the low level of protein is that there is a low concentration of amino acids in the blood. They are almost all taken up by the muscles, where they form some new protein. The liver receives very few of these amino acids and so can no longer synthesise plasma proteins. This affects the water potential of the blood and less of the tissue fluid is reabsorbed. Accumulation of this tissue fluid gives rise to the oedema that is a characteristic of kwashiorkor.

Plasma proteins can also be lost from the body in patients with some forms of kidney disease. In these people there is also a fall in the plasma protein concentration as the plasma proteins escape from the body in the urine, and an accumulation of tissue fluid.

A heart attack or disease of the heart valves can lead to **heart failure**. This is a general term used to describe conditions where the heart is unable to pump out enough blood to meet the demands of the body. It is often associated with complex changes involving the nervous and hormonal systems. Some of these changes result in less sodium being excreted by the kidneys and an increase in the sodium concentration in the blood. Sodium ions pass out of the blood through the capillary walls during the formation of tissue fluid. As a result of the higher sodium concentration in the tissue fluid, there is a smaller water potential gradient between the tissue fluid and the blood plasma. Less tissue fluid is absorbed and again the characteristic swelling caused by oedema will be evident.

Blood and blood cells

We saw in Chapter 1 that living organisms are made from cells. In complex organisms such as animals, similar cells are grouped together to form **tissues**, each of which has a particular function. Epithelial tissue, for example, forms the lining of blood vessels, nervous tissue conducts impulses within the nervous system and muscle tissue is able to contract and bring about movement. An organ consists of a number of different tissues. An **organ** may be defined as a structure that has a particular function in the organism. Arteries, for example are organs. They are made up of different tissues and their function is the transport of blood from the heart. Together organs make up a **system**.

Q 5 An artery is an organ. Name two tissues found in the wall of a large artery.

Blood is a liquid but it is still an example of a type of tissue known as a connective tissue. Other connective tissues, such as those that make up the outer layer of the walls of arteries and veins, have cells which are separated from each other by a **matrix** that does not contain cells. In

Plasma is blood with all the cells removed.

blood this matrix is a liquid, the **plasma**. Within the plasma are the red blood cells, whose main function is the transport of respiratory gases, and the white cells, which are involved in protecting the body from disease. We will look at each of these.

Plasma

If we take a sample of blood and spin it in a centrifuge, the cells will sink to the bottom and leave a clear, straw-coloured liquid at the top. This liquid is the plasma. It contains many of the substances that are transported round the body: glucose and amino acids, hormones, mineral ions and urea. One important feature of plasma is that it is not constant in composition. Figure 6.11 shows the blood system supplying blood to and taking it away from the liver.

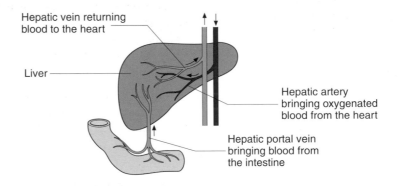

Figure 6.11
The liver receives its blood supply from two sources: the hepatic artery brings oxygenated blood from the heart, while the hepatic portal vein brings blood from the intestine and some other abdominal organs. The blood is returned from the liver to the heart by the hepatic vein.

Immediately after a meal, the blood in the hepatic portal vein has a very high concentration of glucose as it is transporting glucose from the intestine to the liver, where it is stored. The concentration of glucose in this blood vessel will gradually fall as time goes by and the glucose is absorbed. If we eat too much protein, the body cannot store the excess amino acids formed from the digestion of the protein. The liver breaks down the excess amino acids and produces urea. Blood in the hepatic vein has a higher concentration of urea than blood in the other two vessels.

Q 6 In which of the three blood vessels shown in Figure 6.11 would you expect:
(a) the concentration of amino acids to vary most
(b) the concentration of carbon dioxide to be highest?

Red blood cells

The cells that collect at the bottom of the tube when the blood is centrifuged make up what is called the **packed cell volume** (Figure 6.12). This is the percentage of the blood volume taken up by blood cells. In humans, it is approximately 40% and most of the cells involved are red blood cells or **erythrocytes**. Their main function is the transport of respiratory gases.

Erythrocytes (red blood cells) transport respiratory gases.

Plasma contains many of the substances such as glucose and amino acids that are transported by the blood

The packed cell volume: This contains red blood cells and white cells. It makes up about 40% of the total blood volume

Figure 6.12
The cells that collect at the bottom of a tube when a sample of blood is centrifuged form the packed cell volume. It is slightly higher in men than in women.

If you have studied the assignment at the end of Chapter 1, you will already know a lot about red blood cells. They have a number of adaptations that make them very efficient when it comes to absorbing and transporting respiratory gases. These include:

- **Small size:** human red blood cells are only about 7.5 micrometres (μm) in diameter. This is very much smaller than most other cells in the body. A small size means that all of the haemoglobin molecules that these cells contain are close to the surface and this allows oxygen to be picked up and released rapidly.

- **Shape:** a sphere is the shape with the maximum volume. A spherical blood cell can therefore hold a large amount of haemoglobin and transport a lot of oxygen. On the other hand, a blood cell with a flat, disc-like shape would have the maximum surface area, important for the efficient diffusion of oxygen into and out of the cell. Red blood cells are biconcave discs. This is a compromise between having a large volume and a large surface area. It allows the cell to contain a lot of haemoglobin while still allowing efficient diffusion through the large surface area provided by the plasma membrane.

- **Organelles:** red blood cells do not contain either nuclei or mitochondria. This allows more space inside the cell for haemoglobin, the substance responsible for transporting oxygen.

- **Haemoglobin:** haemoglobin is the oxygen-transporting pigment. Much more haemoglobin can be packed into red blood cells than could be dissolved in the plasma. It is also kept in chemical conditions that allow it to function as efficiently as possible in loading and unloading respiratory gases. There would also be a number of disadvantages in having haemoglobin carried in solution in the plasma. These include making the blood so thick that it would have difficulty flowing through the blood system, affecting the water potential and losing the haemoglobin from the body as the blood passed through the kidneys.

Q 7 Explain why red blood cells:
  (a) cannot respire aerobically
  (b) cannot synthesise proteins.

For more about red blood cells, look at the assignment in Chapter 1.

White cells

White cells or leucocytes are usually larger than red blood cells but they are found in much smaller numbers. Some of these cells are able to escape from the blood system and are found elsewhere in the body, so white cell is a more accurate name for them than white blood cell. They are colourless so they have to be stained in order to be seen clearly with a light microscope. Staining allows us to see details such as the shape of the nucleus and the presence of granules in the cytoplasm. Table 6.2 shows the important differences between the main types of white cell.

key term

Leucocytes (white cells) are concerned with protecting the body from disease.

Type of leucocyte	Appearance	Function
Lymphocyte	Has a large, round nucleus and a relatively small amount of cytoplasm	Some types of lymphocyte secrete antibodies Others have different functions, such as killing infected cells and controlling aspects of the immunological process
Monocyte	Has a large, kidney-shaped nucleus	These cells engulf bacteria
Granulocyte	Possess a lobed nucleus and granular cytoplasm	Granulocytes have a number of different functions: some engulf bacteria and others are involved with allergies and in inflammation

Table 6.2

Extension box 3

How long do red blood cells live?

In an adult man there are approximately 5.5 million red blood cells in a single cubic millimetre (mm^3) of blood; in an adult woman the figure is a little less, around 4.7 million per mm^3. With the blood volume of an average adult around 5 litres, this is an enormous number of red blood cells. One interesting fact, however, is that in about 120 days' time, none of the red blood cells that are in your blood now will still be there, they will all have been replaced. So how did we find out how long red blood cells live?

If we want to investigate what happens to a molecule in a biochemical pathway or an individual organism in its natural environment, we need to label it in some way. This principle is also used to find the length of life of a red blood cell. We remove and label some of a person's red blood cells, put them back and then take blood samples at regular intervals until all the labelled cells have disappeared from the blood. A number of modifications of this approach have been tried out.

You will probably know that human blood can be divided into four main groups: A, B, AB and O. Besides this system, however, there are many other ways of grouping blood and these can be used to estimate a red blood cell's life span. The subject of the experiment is given a transfusion of compatible but slightly different blood. Although the red cells introduced will not harm the subject in any way, the slight differences can be detected with suitable antibodies. The number of foreign cells present in the subject's blood can be counted at regular

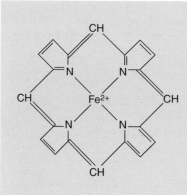

Figure 6.13
The chemical structure of a haem group. Note that it consists of four nitrogen-containing rings grouped round an iron atom.

Figure 6.14
The concentration of ^{15}N in samples of blood from a person who has been fed with ^{15}N-labelled glycine. The upward slope represents the time taken for labelled haemoglobin to be formed and incorporated in red blood cells. The steep fall in the amount of labelled haemoglobin allows us to calculate the length of life of the red blood cells. This ranges from 110 to 140 days.

intervals, allowing us to determine the maximum time for which they can live. This is quite a useful method but it has a problem in that it really only measures for how long one person's blood cells can live in the blood of another person.

If we can use the subject's own blood, we might be able to overcome this problem. A good idea would be to label haemoglobin in some way since this substance is found in large amounts in red blood cells. A molecule of haemoglobin consists of four polypeptide units. Each of these so-called globin chains is associated with an iron-containing haem group. The chemical structure of this haem group is shown in Figure 6.13.

Perhaps the first thing that you might notice is that haem contains iron. Could we label the haemoglobin with a radioactive isotope of iron? Unfortunately this will not work. The red cells disappear from the blood because they are broken down in the liver. The iron is removed from the haemoglobin in these cells and used to make more haemoglobin in more red blood cells so the radioactivity will not disappear from the blood as the labelled cells reach the end of their lives.

We can, however, label the nitrogen in the haem group. We give the subject a small amount of the amino acid glycine containing the radioactive isotope of nitrogen, ^{15}N. We know that the nitrogen in the haem group is derived from glycine so this technique allows us to attach a radioactive label to the red blood cells being formed immediately after the subject was given the glycine. The graph in Figure 6.14 shows the changes in the level of ^{15}N in the blood of one person treated in this way.

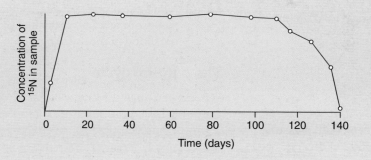

The heart

Humans, like all mammals, have a **double circulation** (see Figure 6.15). The heart has four chambers. Blood flows into the right atrium and out of the right ventricle into the **pulmonary circulation**. Here it is oxygenated as it passes through the lungs. It returns to the left atrium and then goes from the left ventricle into the **systemic circulation**. This is the blood supply to the rest of the body. Most mammals are very active and it is important that oxygen reaches their tissues rapidly. When the blood passes through the many tiny blood vessels in the lungs it loses pressure. By passing through the left side of the heart before being sent round the body, the pressure can be increased, allowing oxygenated blood to reach the organs of the body rapidly.

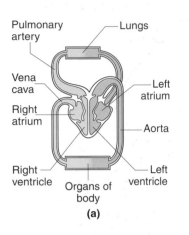

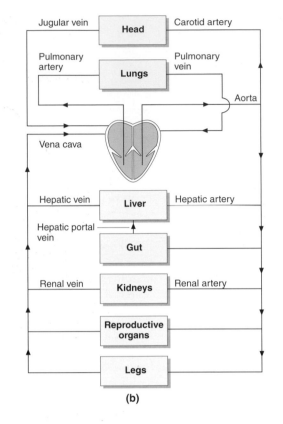

Figure 6.15
(a) The double circulation in a human. Blood is passed from the right side of the heart to the lungs, where it is oxygenated. This oxygenated blood is returned to the heart. It is pumped to the rest of the organs of the body by the left side of the heart.
(b) The main blood vessels that make up the human blood system.

Q 8 Name the blood vessels through which a molecule of urea will pass from where it enters the blood in the liver to where it leaves the blood in the kidney.

The cardiac cycle

When a person is at rest, his or her heart beats approximately 70 times a minute. Each beat of the heart represents a cardiac cycle in which the heart fills and empties. The beating of the two sides of the heart is synchronised. When the left side is filling, so is the right; when the left ventricle contracts, so does the right. The left side of the heart is responsible for pumping the blood all the way round the body; the right side only pumps it to the lungs, which lie relatively close to the heart. The wall of the left ventricle is much thicker than the right and produces much greater pressure when it contracts.

Figure 6.16 shows that, although the cardiac cycle is continuous, we can recognise three main stages: atrial systole, ventricular systole and diastole.

Atrial systole

Some of the blood that enters the atria from the veins flows straight through into the ventricles. The rest is pumped through as the muscle in the walls of the atria contracts. Each atrium is separated from its corresponding ventricle by a valve, the **atrioventricular** (AV) **valve**. The

Q 9 If you could measure a pulse in the pulmonary artery would it beat at the same rate, slower or faster than a pulse in the wrist?

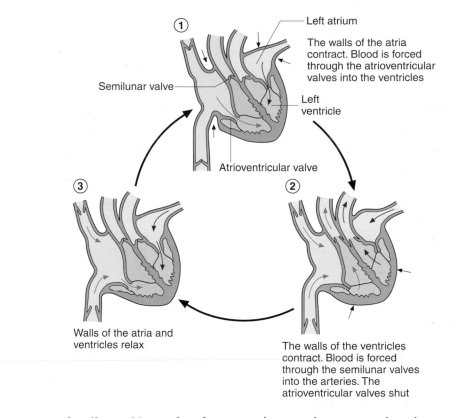

Figure 6.16
There are three main stages in the heart cycle. 1 The atria contract (atrial systole), 2 the ventricles contract (ventricular systole) and finally 3 both atria and ventricles relax (diastole).

① Left atrium — The walls of the atria contract. Blood is forced through the atrioventricular valves into the ventricles
Semilunar valve
Left ventricle
Atrioventricular valve
③ Walls of the atria and ventricles relax
② The walls of the ventricles contract. Blood is forced through the semilunar valves into the arteries. The atrioventricular valves shut

atrial walls are thin, so they do not produce much pressure when they contract but it is enough to open the AV valves and force blood through into the ventricle. Atrial systole takes approximately 0.1 seconds.

Ventricular systole

During ventricular systole, the thick muscular walls of the ventricles contract. The blood in the ventricles is squeezed and its pressure rises. As soon as the pressure of the blood in the ventricles is higher than that in the atria, the flaps of the AV valves are pushed shut. This prevents blood from flowing back into the atria. When a doctor listens to a patient's heart with a stethoscope, the sound made by these valves closing can be heard: it is often described as sounding rather like 'lub'. The pressure of the blood in the ventricles is now higher than that in the arteries leaving the heart. This results in the semilunar valves opening and blood flowing out of the right ventricle into the pulmonary artery, and out of the left ventricle into the aorta. Ventricular systole lasts about 0.3 seconds.

Diastole

The final stage of the cardiac cycle involves relaxation of the atrial and ventricular walls. As the walls of the ventricles relax, the pressure of the blood they contain drops rapidly. The arterial blood pressure is now higher and forces the semilunar valves to shut. This can also be heard with a stethoscope and produces a sound described as 'dub'. With the semilunar valves shut, blood is unable to flow back into the ventricles.

As the muscle in the atrial walls also relaxes, blood is able to flow from the veins into the atria. The walls of the atria now start to contract again, starting another cardiac cycle.

Figure 6.17 summarises the changes in pressure that occur in the left atrium, the left ventricle and the aorta during one complete cardiac cycle. You should be able to explain the main features of these curves. Look first at the curve representing the changes in pressure that take place in the left ventricle. As the muscles in its wall contract, the pressure increases very steeply to reach a maximum value of approximately 16 kPa. When the muscles relax, the pressure falls. There is also a rise in the pressure of the blood in the atria as the atrial walls contract at the beginning of the cycle. The second rise, between about 0.2 and 0.45 seconds, is due to blood flowing into the atria from the veins. Flow of blood from the left ventricle into the aorta results in an increase in the pressure of blood in the aorta. The graph also shows how differences in pressure result in the AV valves and the semilunar valves opening and shutting.

Figure 6.17
Change in pressure during one cardiac cycle. This graph shows figures relating to the left side of the heart. The pressures on the right side are generally a little lower.

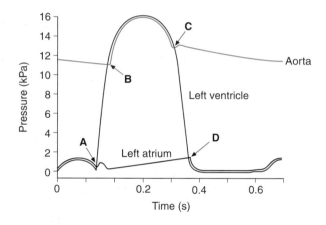

Q 10 At which of the points A, B, C or D in Figure 6.17 would you expect to hear:
(a) the first heart sound (lub)
(b) the second heart sound (dup)?

Coordinating the heartbeat

Suppose you wanted to squeeze all the toothpaste out through the nozzle of a toothpaste tube as quickly as possible. There would be no point in simply squashing the middle of the tube. Some of the toothpaste might come out through the nozzle, but you would be likely to split the tube so that toothpaste leaked out from various places. It is the same with the heart. For the blood to flow from atria to ventricles and then out through the arteries requires the heartbeat to be coordinated. If parts of the heart beat independently, it will not function properly as a pump.

Cardiac or heart muscle differs from other muscles in the body because it is **myogenic**. This means that it beats on its own. It does not need a nerve impulse to make it contract. A heartbeat starts with an electrical signal from an area of muscle in the wall of the right atrium called the **sinoatrial node** (SAN) or pacemaker. The position of the SAN is shown in Figure 6.18.

This electrical signal sets the rate at which the heart beats. Every time the muscle cells in the SAN beat they send out a wave of electrical activity which spreads over the surface of the atria, causing the muscles in the atrial wall to contract. This excitation wave starts at the top of the atria and spreads towards the ventricles, ensuring that the atrial wall contracts in a way that will force blood into the ventricles.

Figure 6.18
Every time the heart beats a wave of electrical activity spreads over the heart from the sinoatrial node. This acts as a signal and causes the heart to beat in a regular, organised way.

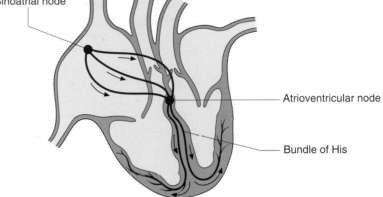

The ventricle muscle must not start contracting until the atrial muscles have finished contracting and squeezed all the blood into the ventricles. The delay in ventricle contraction is brought about by delaying the passage of the excitation wave through to the ventricle. A ring of fibrous tissue between the atria and the ventricles prevents its spread and it can only pass through in one region, the **atrioventricular node** (AVN). After a short delay here, the wave passes down specialised conducting fibres in the wall or septum between the right and left ventricles. These fibres, which form the **bundle of His**, conduct the excitation wave very rapidly to the base of the ventricles. It then spreads upwards through the muscle in the wall of the ventricle. This ensures that the ventricles contract from the base upwards, squeezing blood into the arteries.

Q 11 Explain the importance in the cardiac cycle of:
   (a) delaying the spread of the excitation wave in the atrioventricular node
   (b) conducting the excitation wave very rapidly to the base of the ventricles.

The effect of exercise on the heart

key term

The **cardiac output** is the amount of blood pumped out of the left side of the heart in one minute.

The cardiac output is the product of the **stroke volume**, the amount of blood that the left ventricle pumps out each time it beats, and the **heart rate**, the number of times the heart beats in one minute. This can be represented as a simple equation:

$$\text{cardiac output} = \text{stroke volume} \times \text{heart rate}$$

Q 12 A man's resting cardiac output is $5500\ \text{cm}^3$. His heart is beating at 70 beats per minute. Calculate the volume of blood that his left ventricle pumps out each time it contracts.

If you take some exercise, one of the things you are likely to notice is an immediate increase in heart rate. This is only part of the story, however, as both the rate and force with which the heart beats are continuously adjusted so that cardiac output is matched to the needs of the body. In this way, enough oxygen can be supplied and carbon dioxide can be taken away.

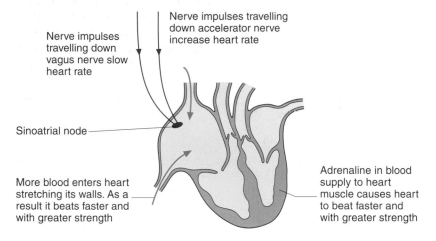

Figure 6.19
Cardiac output can be adjusted to meet the needs of the body. The heart responds to hormones, to electrical impulses passing down nerves from the brain and to changes in the amount of blood returning to the heart through the veins.

Figure 6.19 summarises the three ways in which cardiac output can be adjusted. These involve the heart responding to hormones, nerve impulses from the brain or changes in the volume of blood returning to the heart.

Hormones

When you are under stress or there is a need for action, the hormone adrenaline is secreted by the adrenal glands, which are situated at the top of your kidneys. Adrenaline travels in the blood to all areas of the body. It has many effects and one of them is to increase the heart rate by increasing the rate at which the SAN sends out its waves of electrical activity.

Nerve impulses from the brain

There are two nerves from the cardiovascular centre which go the heart. These are:

- the **accelerator nerve**: impulses travelling down this nerve to the SAN increase the heart rate and impulses going to the heart muscle increase the strength of the contractions

- the **vagus nerve**: impulses travelling down this nerve to the SAN slow down the heart rate.

When you start exercising, the rate of respiration in your muscles increases. More carbon dioxide is produced as a result. Receptors monitor the concentration of carbon dioxide by detecting changes in pH. Nerves from these receptors send impulses to the cardiovascular centre, which responds in an appropriate way. The cardiovascular centre also

key term

The **cardiovascular centre** in the brain is involved in controlling the cardiac output.

responds to other signals. It slows the heart rate in response to high blood pressure and increases it in response to signals from other parts of the brain, such as, for example, in anticipation of activity.

Changes in the volume of blood returning to the heart

During a period of exercise, muscles respire faster and use up more oxygen. A fall in the concentration of oxygen causes the veins bringing blood back to the heart from the body to get wider. More blood therefore enters the heart and stretches its walls more than normal. The heart responds to this by beating faster and with greater strength.

Q 13 **If a person has a heart transplant, the new heart has no nerves going to it. Explain why a transplanted heart can still beat faster if the person exercises.**

Exercise affects the rest of the circulation as well as bringing about an increase in cardiac output. The energy required for increased muscle contraction comes from respiration. More exercise means a faster rate of respiration and the need for more oxygen to be supplied to the muscles. This can be achieved by increasing the amount of blood flowing through the capillaries. When a person is resting, many of the capillaries in muscles are closed. When the muscle is active, most of these capillaries have blood flowing through them. The graph in Figure 6.20 shows the effect of different amounts of exercise on the rate of blood flow through muscle capillaries.

Figure 6.20
This graph shows the effect of exercise on the flow of blood through muscle capillaries. The x-axis shows the oxygen taken up in dm^3 per minute. The more vigorous the exercise, the greater the amount of oxygen consumed.

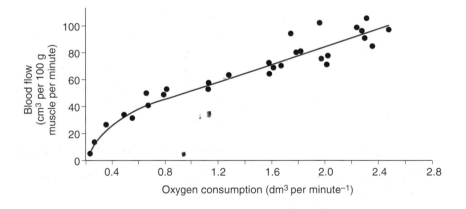

A large increase in blood flowing to one part of the body must be met by a reduction in the amount of blood supplying other parts of the body. During a period of exercise, there is an increase in the amount of blood flowing to the muscles and to the skin but a decrease in the supply to other organs such as the kidneys and those that make up the digestive system. The brain needs a constant supply of oxygen so its blood supply is not affected however severe the exercise.

Summary

- Mass transport in mammals occurs through the arteries, arterioles, and veins.

- Capillaries are important in metabolic exchange and in the formation of tissue fluid.

- Red cells (erythrocytes) in the blood transport oxygen.

- White cells in the blood are involved in protecting the body from disease.

- The heart cycle consists of the atrial systole, the ventricular systole, and the diastole.

- The heart is myogenic. The sinoatrial node (SAN) begins each heartbeat by generating an electrical signal.

- Exercise increases cardiac output and affects the distribution of blood to the different organs in the body.

Assignment

The assignments in the first five chapters in this book have introduced you to some of the skills that a biologist needs. The assignment in this chapter involves using all of these skills. You will need to answer some questions based on a short written passage, apply your knowledge to new situations and handle and interpret data.

Many people living in more developed countries survive to middle and old age. In view of this, it is not surprising that diseases involving the degeneration of organs and systems account for so many deaths. At the top of the list is heart disease, which kills around a third of all people in the UK and leads to the chronic ill-health of many more. One of the most valuable tools we have for studying the heart and detecting heart disease is the electrocardiogram (ECG). Read the following passage, which explains the principles underlying electrocardiography.

The cardiac cycle is controlled by a series of electrical events. A wave of electrical activity spreads over the surface of the heart. The body tissues also conduct electricity, so voltage changes that affect the heart can also be detected on the body surface. Taking an ECG involves attaching a series of electrodes to the surface of the body. They are self-adhesive and coated with a jelly that ensures a good electrical contact. The signals from these electrodes are processed and displayed to give a picture of the electrical events taking place as the wave of electrical activity spreads over the surface of the heart.

Figure 6.21
Recording an ECG. The positions in which the electrodes are attached to the patient are always the same. This enables comparison with an ECG recorded on another occasion.

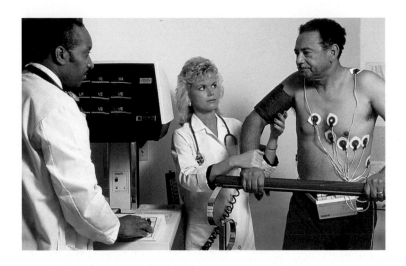

An ECG shows the passage of an electrical signal in a particular direction. If we compare the picture we get from a pair of electrodes, we can see the changes that occur as the electrical activity spreads from one electrode to the other. By attaching the electrodes to different places, we can see what is happening to different parts of the heart. In this assignment we will only be looking at the pattern we see from a single pair of electrodes, showing the spread of electrical activity from the atria to the ventricles.

Figure 6.22 shows an ECG of the events associated with a single heartbeat in a healthy person. Letters are used to identify particular features:

- the P wave shows the spread of electrical activity over the surface of the atria

- the QRS complex shows the spread of electrical activity over the surface of the ventricles

- the T wave shows the electrical recovery of the ventricles; this is the time when the muscle is relaxing and the ventricles are starting to fill with blood.

Figure 6.22
Interpreting an ECG.

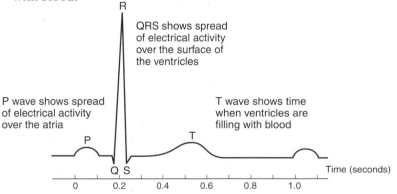

Now use the information in the passage and Figure 6.22 to answer these questions.

1 Describe the path taken by the wave of electrical activity as it spreads over the surface of the heart.

(2 marks)

2 An ECG shows a number of traces which, although they show recordings of the same heartbeats, differ in appearance from each other. Explain why.

(2 marks)

3 Pregnancy has many effects on the body. One of these is to cause the heart to twist slightly so that it is lying in a slightly different position in the chest.
(a) Suggest why a woman's heart may twist and lie in a slightly different position during pregnancy.

(2 marks)

(b) Why might an ECG be misinterpreted if it was not known that the woman from whom it was recorded was pregnant.

(1 mark)

4 Look at the ECG shown in Figure 6.22.
(a) How long does one heartbeat take?

(b) What is this person's heart rate in beats per minute?

(2 marks)

5 Explain why no electrical activity can be detected between the end of the P wave and the beginning of the QRS complex.

(2 marks)

Figure 6.23 shows ECGs obtained from patients whose heartbeats are abnormal. Compare them with Figure 6.22 and answer the question below.

6 (a) Describe how the heartbeats of Patient A differ from those of a healthy person.

(b) Patient B has a condition called ventricular fibrillation. When a patient is being kept in intensive care and ventricular fibrillation is detected, a warning is given and emergency treatment has to be given immediately. Suggest why ventricular fibrillation may result in the death of the patient.

(2 marks)

Look at Figure 6.22 again. The interval between the beginning of the QRS complex and the peak of the T wave is the time when the heart is emptying. The interval between the peak of the T wave and the beginning of the next QRS complex is the time when the heart is filling.

In an investigation, an adult male had his ECG recorded while his heart was beating at different rates. The results are shown in Table 6.3.

Patient **A**

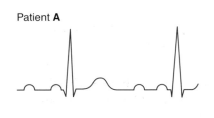

Patient **B**

Figure 6.23

Heart rate (beats minute^{-1})	Emptying time (s)	Filling time (s)
55	0.43	0.66
60	0.43	0.57
70	0.42	0.44
80	0.42	0.33
90	0.43	0.24

Table 6.3

7 (a) Plot these data as a suitable graph using a single pair of axes.

(4 marks)

(b) What do these data show about the way in which heart rate increases?

(2 marks)

Examination questions

1 Table 6.4 shows the pressures measured at two different places in a blood capillary.

	Arterial end of capillary	Venous end of capillary
Blood pressure (kPa)	4.6	1.3
Osmotic pressure of blood plasma (kPa)	3.3	3.3

Table 6.4

(a) Using the data given in the table, explain why fluid leaves a capillary at the arterial but not at the venous end.

(2 marks)

(b) Under normal conditions the total volume of fluid leaving the capillaries is greater than that returned. Explain why this extra fluid does not accumulate in the tissues.

(2 marks)

2 Figure 6.24 shows the changes in pressure which take place in the left side of the heart.

(a) Use the graph to calculate the heart rate in beats per minute. Show your working.

Heart rate = _____ beats per minute

Figure 6.24

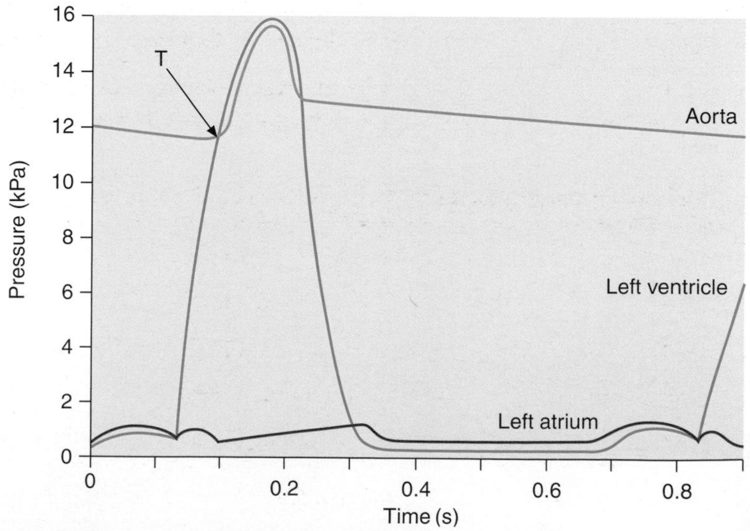

111

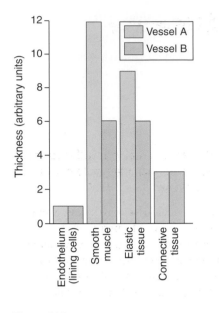

Figure 6.25

(b) (i) Explain, in terms of pressure, why the valve between the left ventricle and the aorta opens at time **T**.

(1 mark)

(ii) For how long is the valve between the left atrium and the left ventricle closed? Explain how you arrived at your answer.

(2 marks)

(c) (i) How would you expect the pressure in the right ventricle to differ from that in the left ventricle?

(1 mark)

(ii) Explain what causes this difference in pressure.

(1 mark)

3 Figure 6.25 shows the relative thickness of parts of the walls of two blood vessels, **A** and **B**. One of these blood vessels was an artery, the other was a vein.

(a) Explain why the thickness of the endothelium is the same for both blood vessels.

(1 mark)

(b) Which blood vessel is the artery? Explain the reasons for your answer.

(2 marks)

(c) Explain how the structure of veins ensures the flow of blood in one direction only.

(2 marks)

Pathogens and Disease

Although we have been fighting a long battle with diseases, they can still surprise us. Diseases like tuberculosis, that we had considered no longer a problem, can suddenly make a comeback, and new diseases can appear.

For instance, in 1994 in Gloucester, there was an outbreak of a frightening disease called necrotising fasciitis. It causes rapid death of tissue which then turns to a liquid mass. Patients die within 24 hours of symptoms appearing. It was reported in newspapers as the 'flesh-eating killer bug'. However, this is not a new pathogen, just a strain of the common bacterium *Streptococcus*. A few cases of this disease are still reported every year. On the other hand, Ebola is a new disease to humans, and so is AIDS.

Ebola virus was first recognised in outbreaks of the disease in 1976 in the African countries of Zaire and western Sudan. These outbreaks resulted in more than 550 reported cases with 340 deaths.

Ebola is a dreadful disease, as was described by the writer Richard Preston in the following way, *"He opens his mouth and gasps into the bag, and the vomiting goes on endlessly. It will not stop, and he keeps bringing up liquid, long after his stomach should have been empty. The bag fills up with a substance known as vomito Negro, or the black vomit. The vomit is not really black; it is a speckled liquid of two colours, black and red, a stew of tarry granules mixed with fresh arterial blood. It is haemorrhagic, and it smells like a slaughterhouse. The black vomit is loaded with the virus. It is highly infective, lethally hot, a liquid that would scare the daylights out of any military biohazard specialist."*

Figure 7.1
An electron micrograph of an Ebola virus.

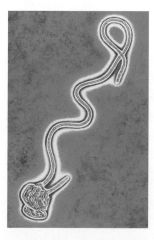

The exact origins and natural host of the Ebola virus remain unknown. However, on the basis of evidence, researchers believe that the virus is animal-borne and is normally maintained in an animal host which is native to the African continent, probably a species of monkey. The virus that was harmless to the monkey is now deadly in its new host, humans.

It is transmitted by direct contact with blood and other secretions of an infected person, usually in the general course of patient care. People can also be exposed to the virus by contact with objects, such as needles that have been contaminated.

The signs and symptoms of Ebola resemble other haemorrhagic fevers beginning with muscle pains and a high temperature. This can lead to breathing problems, bleeding, kidney failure, shock and death. As it has a short incubation period and severe symptoms, those that are infected are recognised and isolated very quickly. Few sufferers, therefore, have the ability to travel far to spread the disease. Western countries, with better hygiene and higher standards of patient care, are unlikely to suffer epidemics of this disease and, as such, it is unlikely to cause a world-wide epidemic.

Microorganisms

The name microorganism is a general description rather than a specific scientific one. It refers to those organisms that are not clearly visible to the naked eye. As such it includes representatives from different groups of organisms – prokaryotes such as bacteria and eukaryoten such as the malarial parasite. Viruses, being even smaller than these organisms, are also included.

Although microorganisms are all small, they have little else in common, being extremely varied in both structure and function. Apart from viruses, all other microorganisms can live and replicate independently of other cells. If the structure and function of microorganisms are diverse, so are their effects. The effects can range from beneficial to pathogenic and they are the cause of many fatal diseases such as tuberculosis, AIDS and malaria. There are many types of microorganism that do not cause disease; some are even beneficial to humans.

Microorganisms and disease

The human body is an ideal incubator for most microorganisms. It provides a warm environment with a constant temperature, a near neutral pH, a constant supply of food and water, a ready supply of oxygen and mechanisms for removing waste materials. Not surprisingly, therefore, our bodies are naturally colonised by a large number of microorganisms. Most bacteria cause you no harm; the few that do are called **pathogens**. The fact that, in most cases, they are harmless is the result of our considerable armoury of physical and chemical defences. The extent to which the pathogen causes damage is known as virulence. For microorganisms to be considered pathogens they must:

- gain entry to the host

- colonise the tissues of the host

- resist the defences of the host

- cause damage to the host tissues.

In 1881, Robert Koch suggested four principles to indicate that a disease is the result of a pathogen. These principles are known as **Koch's postulates**.

- the microorganism must be present in the body of an animal that is suffering from the disease, but it must not be present in healthy animals

- the microorganism can be isolated from the infected host and can be grown in culture

- by introducing some of the cultured microorganism into a healthy host, symptoms of the disease should develop

- the microorganism can be isolated from the new host.

Using these postulates, many microorganisms that are responsible for causing a disease have been identified. However, there are problems, especially when dealing with humans. It would not be ethical to introduce a suspected pathogen into a human to see if the disease developed or not. In addition, some people are resistant to the disease, while some may be carriers. Their bodies contain the pathogen but they never show any symptoms.

How an infection takes hold

Entry

Entry of pathogens occurs broadly in two ways:

Through the skin

The surface of the skin consists of dead, dry cells made of an indigestible protein called **keratin**. This layer, if damaged by cuts and grazes, can form an entry point for bacteria. Mosquitoes, ticks and lice all feed by puncturing the skin and therefore a microorganism can be transmitted into the host blood when they feed.

Through natural openings

The natural openings of the body – the breathing system, the gut, the urinary system and the reproductive system – are lined by thin mucous membranes that create a slightly easier access point for microorganisms. Microorganisms entering the body when air is inhaled, cause diseases such as influenza, TB and bronchitis. Food and water may carry the agents of typhoid, dysentery and cholera into the intestines via the mouth. The genital and urinary opening may allow microorganisms to enter causing cystitis. Those requiring direct contact for their transference are easily transmitted during sexual intercourse. Examples include agents of AIDS, syphilis and gonorrhoea.

Following entry

After infection, a pathogen must do three things to produce a disease:

- attach itself to cells in the host
- penetrate the host cell
- colonise the host tissue.

Attachment

Once a pathogen has entered the body, it needs to fix itself at the site of infection. Most infections begin when the pathogen attaches to the epithelial cells in the digestive, respiratory or urino-genital systems. They are able to achieve this because all cells have plasma membranes that contain different protein molecules. The type of protein present depends on the type of cell. Only if the right protein is present will the microorganism attach.

Microorganisms are able to attach to host cells because molecules known as **ligands**, in the microbial cell wall or outer viral coat, bind with receptor molecules in the host cell membrane (Figure 7.2).

Figure 7.2
Pathogen attaching to host cell.

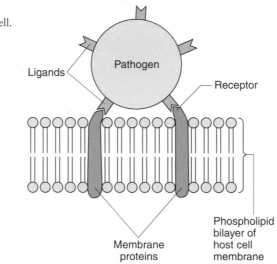

Pathogens with ligands that complement receptors with a specific shape on epithelial tissue, such as the lining of the throat, cannot attach to similar tissues in other animals. The matching is called **host specificity**. The host cell receptors are proteins, and the genes coding for these receptors vary from individual to individual. This means that some people are more susceptible to certain diseases than others.

Penetration

Attachment is sometimes not enough to cause a disease. The ability of the microorganisms to penetrate host cells is also important. In some cases the pathogen produces enzymes that damage the host cell membranes, and allow it to enter. In other cases, the host cell engulfs the pathogen. Phagocytic white blood cells engulf bacteria and try to destroy them but some bacteria can survive this attack because their cell walls are covered in a protective capsule. Once inside the phagocyte, the bacteria, unaffected by their ingestion, are able to reproduce freely.

Colonisation

After attachment and penetration, the pathogen multiplies rapidly and increases in numbers. The need then is to spread the newly formed pathogens throughout the tissue and to other regions of the body. Some microorganisms may produce toxins that cause irritation which leads to responses such as scratching, coughing and sneezing. This helps to spread the pathogen to unaffected areas.

- Enzymes produced by some pathogens enable them to penetrate cells and so gradually invade a tissue

- Microorganisms may enter the lymphatic system via tissue fluid and are carried around the body in this way

The body will fight the infection using a complex system. If the skin and mucus, lining the body, represent the first lines of defence, then the immune responses of the blood and lymphatic system are the second lines of defence. Details of how these operate are given in Chapter 8.

Once bacteria have colonised host tissue, the next stage causes direct damage to cells or indirect damage to cells due to the production of toxins.

Damaging the host

Damage to the host cells

Examples of direct damage to cells include the effects of viral diseases such as AIDS, where the HIV destroys large numbers of lymphocytes. Similarly, malarial symptoms caused by the parasite *Plasmodium* are a result of the loss of red blood cells.

Release of toxins

Many bacteria produce toxins. **Endotoxins** are released by the bacteria as they grow. Endotoxins are substances that form part of the bacterial cell wall, which escape into the body when the bacteria die and the cell wall breaks down.

The extent of any damage the pathogen causes and hence the onset of symptoms is related to the rate at which it multiplies. Pathogens, like those causing gastroenteritis, divide approximately every 30 minutes and so produce symptoms within 24 hours of infection. The bacteria causing leprosy, by contrast, take up to three weeks to divide and so symptoms are not apparent for many weeks. The number of pathogens necessary to cause symptoms also varies. Some pathogens will only cause damage if present in very large numbers, while others such as typhoid bacteria cause harm when their numbers are relatively small.

In summary, the ability of bacteria to cause disease relies on:

- situation – the tissue which the bacteria colonise

- infectivity – how easily a bacterium can enter the host cell and cause infection

- invasiveness – how easily a bacterium or its toxin spreads within the body

- pathogenicity – how poisonous the toxin is.

Organisms that cause disease

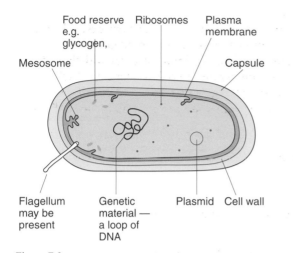

Food reserve e.g. glycogen,   Ribosomes   Plasma membrane

Mesosome   Capsule

Flagellum may be present   Genetic material — a loop of DNA   Plasmid   Cell wall

Figure 7.3
Major structures of a bacterial cell.

Bacteria

Bacteria are single-celled prokaryotic organisms. Their cells are less highly organised than those of eukaryotic organisms such as animals, plants and fungi. Bacterial cells are very small, 0.5–1.0 μm in diameter compared to the average size of a eukaryotic cell of 20 μm.

Bacterial cells have a circular chromosome made of a strand of DNA but may also have one or more smaller circular DNA **plasmids** that can code for antibiotic resistance or toxin production. The structure of a bacterial cell is shown in Figure 7.3.

There are four main types of bacteria, classified by their shape:

Figure 7.4
Four main types of bacteria.

- cocci (spherical)
- bacilli (rod-shaped)

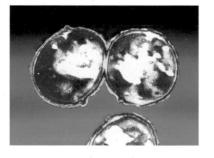

- spirilla (cork-screw)
- vibrio (comma-like)

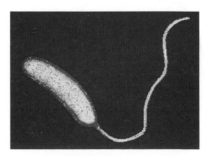

Growth

Bacteria are small, easily dispersed and quickly multiply given a suitable environment.

Each species has its optimum conditions in which it grows best. The life-cycle of a bacterium often involves no more than growth and cell division. Typically, after growing to its maximum size, the rod-shaped *Bacillus subtilis* divides longitudinally into two daughter cells. Cell division begins with replication of the genetic material. The cell then elongates, the genetic material separates and a cross wall forms to divide this elongated cell into two.

Factors affecting growth

Nutrients

Growth depends on both the type of nutrient available and its concentration.

Cells are largely made up of the four chemical elements, carbon, hydrogen, oxygen and nitrogen, with small quantities of phosphorus and sulphur. These six elements are therefore essential for growth. Other elements, e.g. calcium, potassium, magnesium and iron, are no less important and are called macronutrients. Some elements are required in even smaller amounts; these are the micronutrients, e.g. manganese, cobalt and zinc. A further group of substances loosely termed growth factors are also needed. These include vitamins, amino acids, purines and pyrimidines.

Temperature

As all growth is governed by enzymes that operate only within a relatively narrow range of temperature, cells are similarly affected by temperature change. If the temperature falls too low, the rate of enzyme-catalysed reactions becomes too slow to sustain life; if too high, the denaturation of enzymes causes death of the cell. Most microorganisms grow best within a range of 20 to 45°C although some species can grow at temperatures as low as –5°C, and others as high as 90°C.

Q 1 **Explain how storing food in a refrigerator will affect the growth of bacteria on a piece of chicken.**

pH

Microorganisms are able to tolerate a wider range of pH than plant and animal cells, some species growing in an environment as acid as pH 2.5, others in one as alkaline as pH 9.

Oxygen

Many microorganisms are aerobic, requiring molecular oxygen for growth at all times: these are termed obligate aerobes. Some, while growing better in the presence of oxygen, can nonetheless survive in its absence; these are called facultative anaerobes. Others find oxygen toxic and do not grow in its presence; they are the obligate anaerobes.

Q 2 The air in hospital operating theatres and microbiology laboratories is often slightly below atmospheric pressure. How does this reduce the risk of infectious organisms spreading to other parts of the hospital?

Measurement of growth

As bacteria and other microorganisms are so small, it is usual to study the growth of populations rather than individuals. Growth rate is usually measured as the increase in cell number over a period of time (see Box 1).

Box 1

Here is an example showing how to calculate rate of population growth.

The table shows the number of yeast cells in 1 mm^3 of sucrose solution over a 5-hour period.

Time (hours)	Number of yeast cells in 1 mm^3 of sucrose solution
0	1.8×10^3
1	2.1×10^3
2	2.4×10^3
3	2.7×10^3
4	2.9×10^3
5	3.2×10^3

Q 3 Calculate the mean rate of increase in the yeast population between 0 and 3 hours.

$$\text{The rate of population growth} = \frac{\text{Increase in number of yeast cells}}{\text{Time}}$$

$$= \frac{2.7 \times 10^3 - 1.8 \times 10^3}{3 - 0}$$

$$= \frac{0.9 \times 10^3}{3}$$

$$= 0.3 \times 10^3 \text{ cells per hour}$$

Most bacteria multiply by binary fission, which means that one cell divides into two daughter cells. In this way, the cell number doubles every generation. The time taken for a bacterial population to double in number is called the **generation time**. If conditions are ideal, some bacteria can divide every twenty minutes, although this is rare under natural conditions.

Population growth studies are essential in biotechnological industries using microorganisms on a large scale. By studying the growth of microorganisms, the ideal conditions and growth requirements can be found. If microorganisms are cultured in optimum conditions, maximum yield of the product can be obtained.

There are two ways of estimating cell numbers in a microbial population:

- **total cell count** – gives the total number of cells present, whether living or dead

- **viable cell count** – takes into account only living cells since these are the only ones capable of dividing.

Total cell count

The culture, usually in a liquid medium, is sampled at known time intervals and the number of cells in a known volume is counted using a counting chamber such as a **haemocytometer**. This is a large microscope slide with a ruled grid of known dimensions. Looking through the low-power objective lens of the microscope you will see a grid. The type A square in the centre is divided into 25 smaller type B squares. These smaller squares are the ones we normally use for counting, and they are recognisable as they have three lines on each side. A coverslip is placed over the grid, which traps a set depth of 0.1 mm of fluid. The volume of liquid contained within each type A square can now be calculated, as the side lengths are 1 mm. The volume is thus 0.1 mm³ (see Figure 7.5).

Figure 7.5
Diagram to show the three different squares sampled on a haemocytometer. The depth of solution under all squares is 0.1 mm.

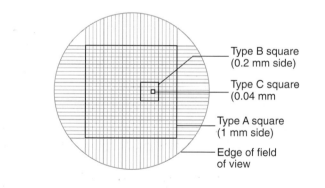

Type B square (0.2 mm side)

Type C square (0.04 mm

Type A square (1 mm side)

Edge of field of view

If a drop of culture containing yeast or bacterial cells is placed on the haemocytometer, the number of cells in a set volume of culture can be calculated. It is not necessary to count all the cells in a type A square; the number of microorganisms in several type B squares is counted to give a representative sample. Five squares are usually enough and generally squares 1, 5, 13, 21 and 25 are used (see Figure 7.6).

A number of different samples are counted and the mean calculated to obtain more reliable results. The number of cells in 1 mm³ or 1 cm³ can be found by simple mathematics.

Figure 7.6
Diagram to show 25 type B squares and which ones are used to sample the population.

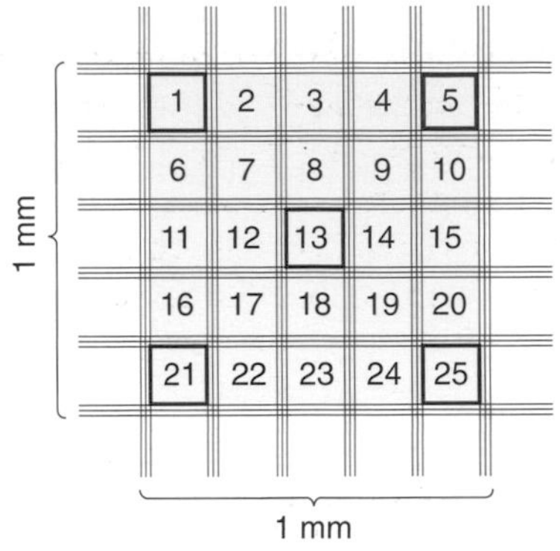

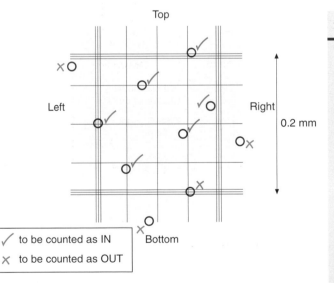

1 mm

1 mm

Counting

Although three lines border the type B square, it is only the **middle** one that is used. There is a rule used to count cells with a haemocytometer, to avoid cells that overlap this line being counted twice. If the cells are touching the middle line of two sides of the type B square then they are counted, if they are touching any line of the other two sides they are not.

With the example in Figure 7.7, let's say that cells in contact with the left and top side of a type B square are to be counted, then those in contact with the right and bottom sides are not counted (see Box 2).

Where the number of cells is too large to count accurately, the broth should first be diluted. The final estimate can then be calculated by multiplying the average count by the dilution factor used.

The haemocytometer provides a **total count** of all cells, whether living or dead.

Figure 7.7
Type B square with a number of yeast cells.

Top

Left

Right

0.2 mm

✓ to be counted as IN

✗ to be counted as OUT

Bottom

Box 2

The sample in Figure 7.7 consists of 6 cells.

This procedure is repeated with four other squares.

Suppose the numbers of cells in this square and four other squares were:

6   8   9   5   7

The sum of the cells in these five squares is **35**.

The mean for one type B square is 7.

Therefore the number of cells in the 25 squares can be estimated as:

$25 \times 7 = 175$

Therefore there are 175 cells in **0.1 mm³**.

In **1 mm³** there will be **1750** cells.

This is our cell count, or population.

Often it is more useful to be able to estimate only the number of living cells in a culture.

Viable cell count

To obtain the number of live cells, a known volume of the original sample is diluted with distilled water and a **serial dilution** is made up. This is done by taking 1 cm³ of the original culture and placing it into 9 cm³ of distilled water in a second tube and then mixing it thoroughly. This solution is now one tenth of the concentration of the original. This process is then repeated (Figure 7.8) providing a range of dilutions.

Figure 7.8
The dilution and plating method for estimating the number of bacteria in a sample of milk.

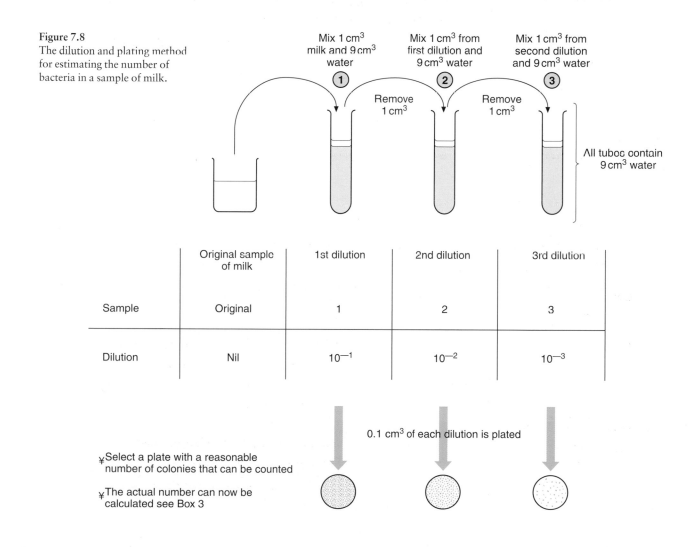

Sample	Original sample of milk	1st dilution	2nd dilution	3rd dilution
Sample	Original	1	2	3
Dilution	Nil	10^{-1}	10^{-2}	10^{-3}

0.1 cm³ of each dilution is plated

¥ Select a plate with a reasonable number of colonies that can be counted

¥ The actual number can now be calculated see Box 3

123

1 cm³ of each dilution is then added to a separate agar plate, spread evenly over it, and left to incubate at a suitable temperature for a few days. Each **living** cell will give rise to a visible colony growing on an agar plate. Some plates will have so many colonies that separate ones cannot be identified; others will have no colonies at all. At one dilution the number of colonies will number 5–50. The actual number is counted and multiplied by the dilution factor to give the number of living cells in 1 cm³ of the original culture. This is called a **viable count** (see Box 3).

Box 3

Suppose that the number of colonies on the 10^{-3} dilution plate is **35**.

- Remember only 0.1 cm³ of the dilution was spread onto the plate.
- Therefore the number of viable colonies in 1 cm³ of 10^{-4} dilution of milk

$$= 35 \times \frac{1}{0.1}$$

$$= 350$$

- Remember the sample was diluted by 10^{-3}
- Therefore the viable number of bacteria in 1 cm³ of the original sample is

$$= 350 \times 10^3$$

$$= 3.5 \times 10^5 \text{ (or } 350\,000)$$

Q   4   The numbers of yeast cells present in a 1 cm³ culture were found to be:

- 2.5×10^4 cells per cm³ using a haemocytometer

- 1.8×10^3 cells per cm³ using serial dilution and pouring plates.

  Suggest why the results obtained are different.

Aseptic conditions

Pure cultures of a single type of microorganism need to be grown free from contamination with others. A number of techniques are used to sterilise equipment, instruments, media and other materials to prevent contamination. High temperatures and disinfectants are effective ways of sterilising instruments, working areas and culture media.

Growth patterns of bacteria

When an organism is introduced and successfully establishes itself in a new area, its population grows slowly at first. It then undergoes a period of increasingly rapid population growth before reaching an equilibrium wherein its numbers remain more or less constant.

A typical growth curve, showing the changes that occur in a population of bacteria, is illustrated in Figure 7.9 and consists of four phases.

Figure 7.9
A population growth curve for bacteria.

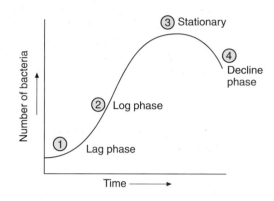

Lag phase

This is the period directly after the cells have been added to the nutrient medium. The increase in the population of the bacteria during this time is slow. They increase in size but only very slowly in number. It is a period of adjustment to new conditions. During this stage the bacteria start to produce the enzymes needed to utilise the nutrient medium.

Log phase

During this period the supply of nutrients, oxygen and other factors affecting growth are in plentiful supply. Waste products are being produced, but they have not built up to harmful levels. The rate of cell division is at its maximum with the population doubling and redoubling. This is **exponential** growth.

Stationary phase

During this phase the population has reached equilibrium. The number of new cells added to the population is more or less balanced by the number of cells that are dying. If nothing is added to the culture, the nutrients are used up. Toxic waste products accumulate, and become limiting.

Death/decline phase

While the total number of cells remains constant, the number of living cells starts to decrease. An ever-increasing number die from lack of nutrients or because of poisoning by a build-up of toxic waste products.

Infections caused by bacteria

Tuberculosis

Mycobacterium tuberculosis is the bacterium that causes TB. It may affect almost any organ, although the lung is the most common. Once a widespread disease in the UK with up to 40 000 cases a year, it is now relatively rare in developed countries due to improved living standards and the use of antibiotics and vaccination. These have made TB treatable and preventable, but this disease remains a major killer elsewhere. Significantly there has been a worrying increase in the disease both in the UK and the USA. This is due to the appearance of drug-resistant strains of *Mycobacterium tuberculosis* and to the decreasing number of people being immunised.

Q **5** **Suggest a reason why infectious diseases, such as TB, are now increasing.**

The most common route of infection is from breathing in droplets exhaled during coughing by an infected person.

Initial infection is termed **primary TB**. The symptoms can be a fever, loss of weight and a persistent cough but often there are no symptoms at all. Some bacteria may remain dormant for up to 30 years and then re-emerge as **post-primary TB** in the lungs or elsewhere. Without treatment, the bacteria destroy the lung tissue and cause accumulation of fluid in the pleural cavity between the lungs and the chest wall. Infected individuals might cough up blood because the bacteria destroy the lung tissue. A chest X-ray shows the damage as distinctive pale patches often referred to as a shadow (Figure 7.10).

Treatment involves the administration of antibiotics for a period of six months to ensure complete eradication. Combinations of drugs are used as this discourages the development of strains that are resistant.

The vaccine is a live attenuated strain of *Mycobacterium tuberculosis*. Although this BCG vaccination is very useful in preventing certain types of TB, there is no completely effective vaccine against TB of the lungs.

Figure 7.10
Chest X-ray showing a lung with the distinctive paler patches of tuberculosis.

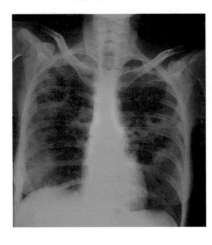

Salmonellosis

The group of bacteria known as *Salmonella* contain many different species of bacteria, which can cause a variety of diseases including typhoid, paratyphoid fevers and some forms of food poisoning.

When contaminated food is eaten, the bacteria pass into the intestine. Here they enter the cells lining the small intestine and multiply. As the population increases, some bacteria die and release endotoxin which irritates the lining of the intestine and causes the typical symptoms of food poisoning.

Food poisoning caused by *Salmonella* is characterised by diarrhoea, abdominal pain and vomiting. The incubation period can be as short as 12 hours, but the symptoms normally subside after 2–3 days and recovery is complete. In a few cases, dehydration or blood poisoning may arise and can, on rare occasions, lead to death. The dehydration can be treated using oral rehydration therapy (ORT), a mixture of glucose and salts in water. In severe cases it may be necessary to replace body fluids with an intravenous drip. The very young, the elderly or those with reduced immunity are most at risk.

Q 6 How does oral rehydration therapy help patients suffering from diarrhoea?

The best way of dealing with *Salmonella* is to avoid getting it in the first place. Careful attention to basic food hygiene removes most of the risk.

Viruses

Because viruses have no cellular structure, they don't respire or feed. They can only reproduce when inside other living cells. They are about 50 times smaller than bacteria, and therefore cannot be seen with a light microscope. They have been recognised as infective agents since as early as 1892 but their structure was determined only by extensive studies using the electron microscope.

General structure

Viruses have a very simple structure consisting of a core containing genetic material of DNA or RNA. This is surrounded by a protective coat of protein called a **capsid**, made of subunits called **capsomeres**. Some viruses also have an envelope of lipoprotein around the capsid (Figure 7.11).

Figure 7.11
General structure of a virus.

Viruses harm the cells and tissues of their host because, in order to multiply, a virus must take over the host cell. They have no organelles and thus use the host cell's organelles to make more virus particles. Viruses cause a wide range of diseases including influenza and AIDS.

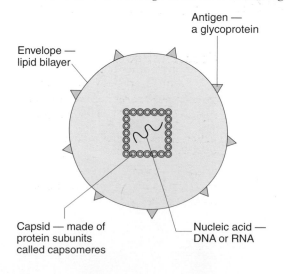

Antigen — a glycoprotein

Envelope — lipid bilayer

Capsid — made of protein subunits called capsomeres

Nucleic acid — DNA or RNA

Figure 7.12
The structure of a Human Immunodeficiency Virus.

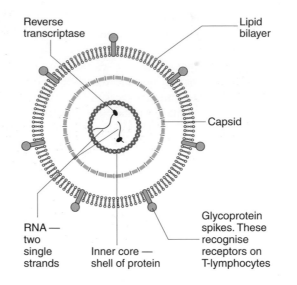

Human Immunodeficiency Virus (HIV)

HIV belongs to a group of viruses called **retroviruses**. Retroviruses have some unique features:

- they have RNA as their genetic material, not DNA

- they contain the enzyme **reverse transcriptase** which enters the host cell along with the RNA. This allows the cell to make viral DNA from viral RNA

- when the viral DNA is inserted into the host cell's DNA it forms a **provirus.**

The viral infection can stay dormant for some time, even several years, before the provirus starts to make new viral RNA molecules and therefore further virus particles. However, during this dormant period, every time the host cell divides the provirus is replicated too, so the numbers of infected cells increase.

Various retroviruses have been discovered in humans and other animals, but the best known is HIV, the virus that causes **Acquired Immunodeficiency Syndrome (AIDS)**. HIV follows a similar pattern of infection to other retroviruses (Figure 7.13).

Figure 7.13
Simplified diagram to show HIV's cycle of infection.

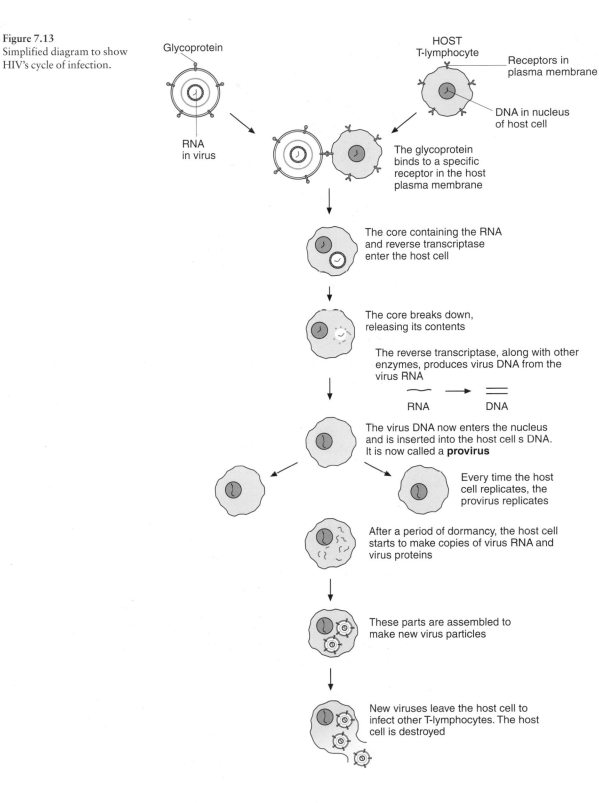

Glycoprotein

RNA in virus

HOST
T-lymphocyte

Receptors in plasma membrane

DNA in nucleus of host cell

The glycoprotein binds to a specific receptor in the host plasma membrane

The core containing the RNA and reverse transcriptase enter the host cell

The core breaks down, releasing its contents

The reverse transcriptase, along with other enzymes, produces virus DNA from the virus RNA

RNA → DNA

The virus DNA now enters the nucleus and is inserted into the host cell s DNA. It is now called a **provirus**

Every time the host cell replicates, the provirus replicates

After a period of dormancy, the host cell starts to make copies of virus RNA and virus proteins

These parts are assembled to make new virus particles

New viruses leave the host cell to infect other T-lymphocytes. The host cell is destroyed

HIV infects T-helper lymphocytes and other white cells called macrophages and destroys them. An unusual feature of HIV is that there is considerable variation between viruses isolated from different people. This is why a vaccination against the disease has been very difficult to develop.

HIV is transmitted in three main ways:

- sexual contact with an infected person

- an infected woman passing it on to her baby through the placenta

- blood from an infected person entering another person's bloodstream, either through sharing needles, receiving infected blood during blood transfusions or through a wound.

Signs and symptoms

While in the provirus state, the virus is largely hidden from the immune system. This stage can last from 2 to 10 years, during which time the person has no symptoms, appears healthy and feels well but there will be antibodies to the HIV in the blood. These antibodies can be identified and the person is said to be HIV positive. After this dormant phase, the number of T-helper lymphocytes in the body falls dramatically as they are destroyed. This marks the beginning of AIDS. The immune system becomes less and less effective. The infected person is said to be immunodeficient. Patients with AIDS frequently become infected by other bacteria and viruses. This leads to diseases such as pneumonia, tuberculosis or **Kaposi's sarcoma,** a form of cancer. Any of these may result in death.

Figure 7.14
Photograph showing a person with Kaposi's sarcoma.

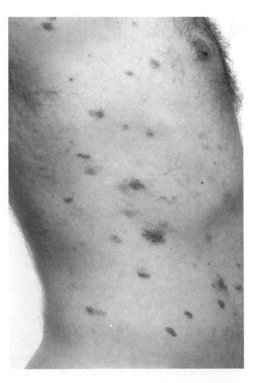

Treatment and control

Researchers are trying to develop drugs against AIDS. **AZT** prevents reverse transcriptase making DNA from RNA. Other drugs block the receptors on the host cell, preventing the virus from entering. These drugs do not cure the condition; they only prolong and improve the quality of life of AIDS patients. An effective vaccine is not yet possible as the HIV can change its surface proteins and evade the immune system. Therefore the priority is to limit its spread.

Parasites and disease

Living organisms interact with one another in different ways. These interactions may involve benefits to both of the organisms concerned or there may be an advantage to just one of them.

A parasite may be defined as an organism that lives in or on a host organism. The parasite gains a nutritional advantage by getting nutrients from the host. In this relationship the host is clearly at a disadvantage and is harmed in some way. It is often difficult to distinguish between parasitism and predation. However, with parasitism, the parasite spends a significant length of time feeding on the host. Another difference is that the host can produce an immune response to a parasite but not to a predator. A parasite's success may be measured by its ability to resist this immune reaction.

There are two main categories of parasites:

- endoparasites, living inside the body of the host

- ectoparasites, living on the outside of the host.

In both cases the parasite needs to be able to:

- maintain its position in or on the host

- resist the host's attempts to destroy it

- reproduce so that the offspring are able to find a suitable habitat in which to develop.

Many different species live as parasites but they have similar adaptations to their way of life. These include some but not necessarily all of the following features:

1   Ability to penetrate the host

Some parasites have structures which enable them to enter into the body or cells of the host.

2   Means of attachment to the host

Many parasites have attachment structures enabling them to cling to the host. Adult blood flukes, *Schistosoma*, are worms that live in veins near the bladder. They are small and therefore could be easily moved by the blood. Suckers help them attach to the walls of the veins.

The malarial parasite *Plasmodium* spends much of its adult life inside the cells of its host. Attachment is therefore not a problem.

3 Various ways of avoiding or overcoming the host defence mechanism

The blood fluke *Schistosoma* synthesises substances that switch off the host's immune system: the parasite coats itself with molecules from the host's red blood cells and that the host mistakenly recognises those cells as its own.

When the malarial parasite is in the blood it will be attacked by the white cells of the immune system. However, within an hour the parasites are inside the cells of the liver and are protected. Also, antigens on the surface of the malarial parasite change frequently and the immune system cannot keep up.

4 Development of reproductive organs

The chances of a parasite successfully completing its life-cycle and infecting a new host are remote. By producing large numbers of offspring, these chances are improved. The reproductive powers of many parasites are phenomenal. In the organism causing malaria, reproduction occurs during four stages in its life-cycle, two of them in the human host and two in the mosquito. In the type of malaria caused by *Plasmodium vivax*, the number of parasites may exceed 30 000 per mm^3 of blood. Again with schistosomiasis the process is similar and each egg laid by an adult worm can produce up to 200 000 cercariae larva.

5 A complex life-cycle

This again improves the chances of a parasite infecting a new host. One of the greatest problems facing any parasite is getting from one host to another. Many parasites have a secondary or intermediate host that conveys the parasite from one primary host to another. Thus the *Anopheles* mosquito transfers the malarial parasite *Plasmodium* from one person to another. An aquatic snail transfers *Schistosoma*. Organisms, such as the mosquito and the snail, which carry a parasite from one primary host to another, are called **vectors**.

Endoparasites with a primary and secondary host may have a number of structurally distinct larval stages, each suited to a different environment and each allowing rapid asexual reproduction.

6 The reduction of body systems other than those involved with reproduction

Many endoparasites show structural degeneration, or even total loss, of certain organs. They may lack sense organs and frequently have a reduced nervous system. Gut parasites like the tapeworm lack an alimentary canal, as they have no need for a digestive system. They only need to absorb the products of their host's digestion. Most of the malarial parasite's life is spent inside one of its host's red blood cells. It does not have to move to find food and the medium in which it lives has the same water potential as its own cytoplasm. It has lost the capacity to move and to regulate its cell water content.

Q 7 The cercariae larva of *Schistosoma* emerge from water snails during the daytime. How is this behaviour an advantage to the parasite?

Schistosoma does not have a complex nervous system. This can be linked with the fact that conditions within the host are constant and the parasite does not have a locomotory system.

When you think about it, it is not in the interests of a parasite to cause too much damage to its host. Suppose that a parasite killed its host in a short time. In that case the parasite would either perish with the host or be forced to begin the hazardous business of finding a new host.

Extension box 1

Plasmodium (malarial parasite)

Plasmodium is the parasite that causes malaria. This is a very serious disease which causes more human deaths in the tropics than any other organism. The World Health Organisation estimates that each year, 270 million new malaria infections and 2.7 million deaths occur worldwide. The malarial parasite has a very complex life-cycle involving asexual stages in the liver and red blood cells of humans and a sexual stage which begins in humans and continues in *Anopheles* mosquitoes.

Plasmodium has two different hosts during its life-cycle, humans and mosquitoes. The female mosquito sucks blood prior to laying eggs and this enables her to transmit malarial parasites from one person to another.

The life-cycle is complex, with many stages (see Figure 7.15). The parasite initially invades the liver where it multiplies before releasing vast numbers of new parasites into the bloodstream. Here they invade red blood cells. Within these cells they multiply again, destroying the red blood cells. At this stage the parasite develops into male and female stages that remain in the blood until taken up by a feeding female mosquito.

Figure 7.15
Life-cycle of *Plasmodium*.

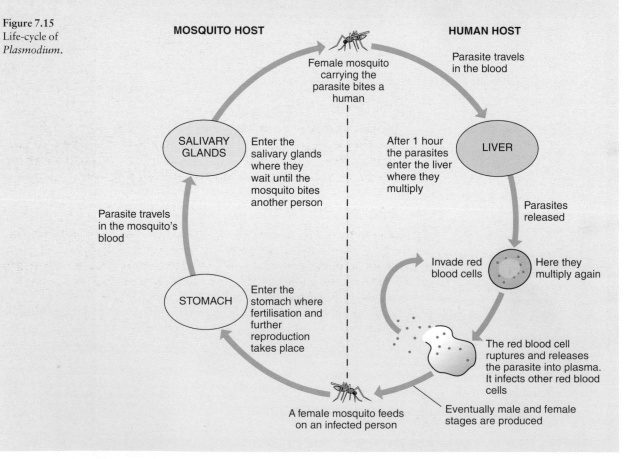

134

In the stomach of the mosquito they mature and fertilisation occurs. The resultant zygote burrows through the stomach wall and forms cysts. Large numbers of a further stage are released from this cyst into the mosquito's blood from where they migrate to the salivary glands to be injected into the next human host.

Signs and symptoms

One of the most characteristic features of malaria is the pattern of fever (Figure 7.16), with temperature peaks at regular intervals which correspond to a new generation of parasites bursting out of the red blood cells. The red blood cells are destroyed causing anaemia. The parasites can also block capillaries, causing pain and damage as they limit blood supply to organs in the body.

Figure 7.16
Diagram to show the cyclical pattern of chills and fevers associated with the life-cycle of *Plasmodium*. Note that the temperature peaks correspond to the bursting of thousands of red blood cells.

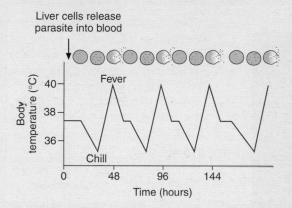

Treatment and control

The best method of control is to break the cycle by destroying the secondary host, the mosquito. Many methods are used. These include spraying the breeding grounds of stagnant water with oil to drown the air-breathing larva, and using insecticide to kill the adults.

Anti-malarial drugs

Chloroquine is the most commonly used anti-malarial drug. It lowers fever and reduces the number of parasites in the body. However, resistance to Chloroquine is widespread and therefore it is now used in a combination with other drugs. Despite decades of research, no-one yet has managed to develop a vaccine that is effective.

Extension box 2

Schistosoma (Blood fluke)

Schistosomiasis, also known as **bilharzia**, results from infection by a blood fluke called *Schistosoma* (Figure 7.17). There are three different species of schistosomes, *S. mansoni*, *S. haematobus* and *S. japonicum* but all these parasites have a similar life-cycle.

Schistosomiasis affects over 200 million people in the world, mainly in the tropics. Of these, 800 000 die each year and 120 million have severe symptoms.

Life-cycle

The adult male and female flukes live in the blood vessels of the abdomen and pelvis of humans. The male is smaller than the female and he is permanently held in a groove in her body. The fertilised eggs are deposited in the blood vessels of the host. The huge numbers of eggs deposited cause the vessels to rupture, and discharge the eggs into the intestine. From here they are carried to the exterior in the faeces. The main damage to the body is caused by the eggs working their way into adjacent organs such as the bladder and large intestine in order to reach the outside of the body. The presence of multitudes of spiny eggs in human tissues causes severe inflammation, abscesses, scarring, and pain on passing urine which often contains blood. Heavy blood loss can lead to anaemia and death.

Not all of the eggs reach the outside of the body; some of them drain with the blood into the larger blood vessels of the abdomen and enter the liver and spleen, where they may cause severe damage.

The eggs gradually work their way through the walls of the blood vessels into the bladder or large intestine and are excreted by the host in the urine or faeces. If there is a modern sewage system, that is the end of the story. However, in some countries people often urinate in ponds and streams, and human faeces are regularly used as fertiliser. Thus the eggs get into water in rice fields, irrigation canals or rivers, where they hatch into tiny ciliated larvae, the **miracidia**. This lives only for a few hours and dies unless it reaches the intermediate host, which is an aquatic snail. If the miracidium larva reaches the snail, it attaches to the skin and digests its way through into the bloodstream. Here it grows to form a structure called a **sporocyst**. Each sporocyst produces hundreds of further daughter sporocysts, and each of these in turn produce many tiny **cercariae** larva that leave the snail in search of the primary host, man. This multiplication of the parasite in the body of the snail is an important feature of the life-cycle, for it greatly increases the chances of ultimately infecting a human.

The tiny cercariae have forked tails and are strong swimmers. If a person comes into contact with water containing these creatures, they bore into the skin or membranes of the mouth and reach the bloodstream. They are then transported around the body in the blood and settle in the vessels of the abdomen or pelvis, where they mature into adult male and female worms (Figure 7.18).

Figure 7.17
Schistosoma has the ability to reproduce in huge numbers. The male lies permanently attached to the female, fertilising eggs that will eventually pass out of the body.

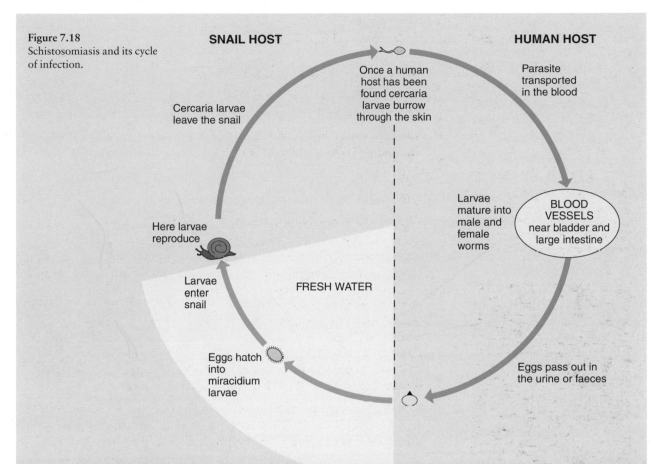

Figure 7.18
Schistosomiasis and its cycle of infection.

SNAIL HOST

HUMAN HOST

Once a human host has been found cercaria larvae burrow through the skin

Cercaria larvae leave the snail

Parasite transported in the blood

Larvae mature into male and female worms

BLOOD VESSELS near bladder and large intestine

Here larvae reproduce

Larvae enter snail

FRESH WATER

Eggs hatch into miracidium larvae

Eggs pass out in the urine or faeces

Signs and symptoms

The body responds to an infection by trying to 'wall off' the eggs that are produced by the worms within the veins producing scar tissue. Scar tissue can also cause the liver to enlarge or can lead to blockages in the bladder and the kidney.

Treatment and control

We fight the disease by using chemicals to destroy the secondary host (the snail) but this is difficult as the substances used are harmful to the environment. People are warned to avoid infected water. Once infected they can be treated effectively with drugs, but only if taken in time. However, this does not prevent reinfection and only continued, repeated treatments will keep the disease under control.

Summary

This chapter should help you be able to:

- understand the sigmoid growth curve of a bacterial population and be able to explain the characteristic log, lag, stationary and decline phases

- describe ways of investigating population growth of bacteria or yeast involving sterile technique and the use of a haemocytometer

- describe the structure of the human immunodeficiency virus and the way it replicates

- describe Koch's postulates

- understand that disease can result from pathogenic organisms penetrating any of the body's interfaces from the environment with reference to *Salmonella spp*, *Mycobacterium tuberculosis* and HIV

- describe the way that microorganisms can cause disease by damaging the cells of the host with reference to *Salmonella spp*, *Mycobacterium tuberculosis* and HIV

- explain the principal adaptations of parasites to their way of life as illustrated by *Plasmodium* and *Schistosoma* to illustrate:

  - their ability to survive in the hostile environment within the host

  - the reduction of locomotory and other structures

  - the modification of reproduction and the way that the life-cycle is associated with infecting a new host.

  - the adaptation of the reproductive system and the life-cycle that results in infection of new hosts.

Figure 7.19 The skin forms a layer which normally protects the body from harmful microorganisms such as bacteria. This layer is damaged in patients with burns. The healing tissue is vulnerable to infection by bacteria such as *P. aeruginosa*.

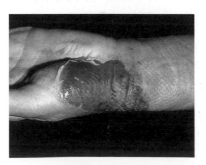

Assignment

Getting a hold

Pseudomonas aeruginosa is a species of bacterium. It doesn't usually cause problems for healthy people but it is responsible for severe lung infections in patients suffering from AIDS or cystic fibrosis (see Chapter 11). It also colonises exposed tissue resulting from burns (Figure 7.19).

Figure 7.20
P. aeruginosa

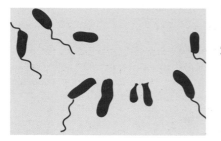

The drawing in Figure 7.20 was made from a photograph showing *Pseudomonas aeruginosa* bacteria.

1. Explain why these bacteria may be described as flagellate bacilli.

(2 marks)

2. (a) Use a ruler to measure the lengths of the bacteria in this drawing. Measure only the cell body in each case. Use your measurements to calculate the mean length in millimetres.

(1 mark)

   (b) The magnification of this drawing is × 2000. Use the answer from part (a) to calculate the actual mean length of these bacteria. Give your answer in micrometres (μm) and show your working.

(2 marks)

One of the things you will need to do as you progress with your study of human biology is to read and understand material from different sources. The second part of this assignment is based on a passage describing how *P. aeruginosa* gets established in its human host. Read the passage then answer the questions which follow. They should help you to understand the main ideas.

Let's consider what happens when the bacterium *P. aeruginosa* lands on tissue damaged as the result of a burn. At first there will only be a few bacteria present. They will do little harm as they grow and divide and increase in number. There comes a point, however, when they behave differently. Once they have reached a critical population density, they start to produce a range of chemical substances. These include toxins which break down the proteins making up the host's cells. The infection is now much more serious. As the cells are destroyed by these toxins, the bacteria spread into the body and bring about serious illness.

How is this change in behaviour brought about? The answer is by something called quorum sensing. Bacteria such as *P. aeruginosa* produce substances which act as chemical signals. These substances are small molecules which diffuse from the bacterial cell into the immediate surroundings. As the number of bacteria increase so does the concentration of this messenger substance. Another thing about the messenger molecules is that they are very specific; each species of bacterium produces and can detect its own messenger molecules. What happens then is that once the bacterial population reaches a critical density, the concentration of messenger molecules around them will be high enough to trigger a change in behaviour. All the bacteria will start to produce toxins at the same time.

As biologists, it is important that we use the correct scientific terms to describe our subject. When we write about microorganisms, for example, we shouldn't use words like "germ" and "bug". We will look at some of the scientific terms in the passage in the next question.

3. (a) The toxins produced by *P. aeruginosa* are sometimes referred to as virulence determinants. Use information on page 115 to explain why.

(2 marks)

(b) If you look up the word 'quorum' in a dictionary, you will see it described as the number of people who must be present before a meeting is valid. Use this definition and information in the passage to explain what is meant by the term quorum sensing.

(2 marks)

(c) The passage refers to the bacteria reaching a critical population density. In what units would you express the population density of bacteria?

(1 mark)

The next two questions are intended to help you to understand the main ideas in the passage. They should also help you to appreciate why the features that have been described are important to the bacteria.

4. Sketch graphs to show:

(a) the change in the population density of the bacteria after the initial infection;

(2 marks)

(b) the relationship between the population density of the bacteria and the concentration of the messenger substance.

(2 marks)

5. (a) As you read earlier in this unit, harmful bacteria must gain entry into the host and resist its defences. Suggest the advantage to the bacteria of not producing toxins when small numbers of bacteria are present.

(1 mark)

(b) How may the production of toxins by the bacteria help them to colonise the tissues of the host?

(2 marks)

6. Now that you have answered the questions above, write a short summary of the main points in the passage. Where possible use your own words and write concisely. You may find it useful to set out your summary as a series of bullet points.

Finally, we will try to take the idea in the passage a little further. Doctors treat many bacterial diseases with antibiotics. Unfortunately many bacteria, *P. aeruginosa* included, are becoming resistant to these antibiotics. Drug companies and research biologists are constantly looking for new ways of treating diseases caused by bacteria.

7. Suggest how the information given in this passage might be useful in the hunt for new ways to treat diseases caused by bacteria.

(3 marks)

Examination questions

Schistosoma is a platyhelminth parasite. Part of its life-cycle is spent in a human host and part in a freshwater snail. The life-cycle is summarised in Figure 7.21.

Figure 7.21

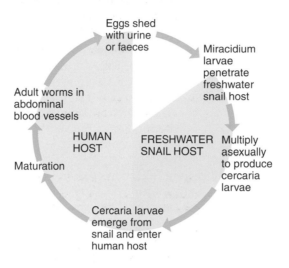

Suggest how each of the following features is an adaptation to the animal's way of life:

1 (a) Adult worms respire anaerobically.

(2 marks)

(b) Hatching of the eggs is stimulated by a less negative water potential.

(2 marks)

(c) Cercaria larvae move towards warm objects.

(2 marks)

2 Figure 7.22 shows the structure of a human immunodeficiency virus (HIV).

Figure 7.22

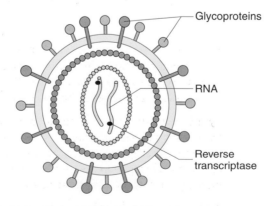

(a) The virus attaches to the plasma membrane of a helper T-lymphocyte. The viral RNA and the reverse transcriptase then enter the lymphocyte. What is the function of each of the following?

   (i) the RNA; *(1 mark)*

   (ii) reverse transcriptase? *(1 mark)*

(b) One form of pneumonia and some types of cancer are normally very rare. They are much more common in people who are infected with HIV. Explain why.

(2 marks)

(c) Viruses such as the influenza virus and HIV have very high mutation rates. Suggest why this makes it difficult to produce a vaccine against them.

(2 marks)

3 Figure 7.23 shows part of a haemocytometer grid viewed at a magnification of ×700. The triple-line square measures 0.2 mm × 0.2 mm. The chamber contains a yeast suspension at a depth of 0.1 mm.

Figure 7.23

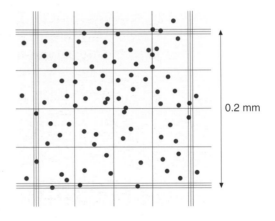

0.2 mm

(a) Assuming that this triple-lined square is typical of all the squares on the haemocytometer grid, how many yeast cells are there in 1 mm³ of the solution? Show your working.

(3 marks)

(b) The dilution factor of the sample counted was × 10⁻³. How many yeast cells were there in 1 mm³ of the original suspension?

(1 mark)

Immunology

In recent years medical advances have highlighted the immune system in a new way. The development of the transplantation of organs and tissues from one person to another has led to a lot of research into the immune system and its suppression. However closely the tissues of a donor organ are matched to the tissues of the patient who is to receive the transplant, a perfect match is never possible unless the donor is an identical twin. This means that, to a greater or lesser extent, the immune system of the recipient will set out to destroy or reject the donated organ.

The problem is how to strike a balance to prevent the recipient from rejecting the transplanted organ without reducing the action of the immune system to the extent that the patient dies from a succession of infections against which the body cannot respond. A cocktail of immunosuppressant drugs prevents rejection without totally disabling the immune system. Transplant patients have to take these drugs for the rest of their lives.

Figure 8.1
Donor card showing consent that all or part of a person's body is available for medical use. However, not enough people carry these cards and there is now a shortage of donor organs for transplantation. Could pigs supply organs for transplanting into humans?

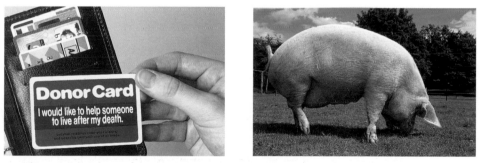

However, there are simply not enough organs to give to all those who would benefit from a transplant; the waiting lists are growing longer and fewer than half of those waiting receive the transplant they need. Attempts have been made to increase the number of donors, including educating the public, changing the law to favour donation, using living donors of kidneys and liver or lung lobes, and using organs removed from victims at accident scenes. These measures still fall short of solving the shortage.

One option is to use animal organs in humans. Some animal tissues are already used as heart valve replacements or as temporary skin dressings. Diabetic patients have been treated for years with pig insulin and some diabetics have received transplants of pig pancreatic islets, the structures that make insulin. Should successful transplants from other animals become a reality new treatments would be available for patients with a range of serious diseases.

Until recently, the transplantation of organs from other animals seemed unlikely because of the high levels of drugs required to prevent rejection. These substances are toxic and depress the immune system so much that the patient is highly susceptible to infection. Now, however, we can modify the donor animals genetically to reduce the rejection response. Based on current evidence, the pig has been selected to be developed as a donor species. Pigs can be bred in sterile conditions and screened for pathogenic microorganisms.

There is still a large amount we do not know about the rejection of these grafts, and we are not certain which combinations of immunosuppressive drugs to use to prevent rejection without killing the patient. A careful risk analysis must be carried out. Most scientists seem to think that the risk is very low, but not zero.

One of the most pressing needs for humans is to avoid being attacked from the inside by microorganisms such as bacteria and viruses. Although many microorganisms do little harm to their host, some cause disease and are called **pathogens**.

Q 1 What conditions inside our bodies make it ideal for the growth of microorganisms?

The human body has three lines of defence against the invasion of pathogens:

- barriers preventing entry
- a non-specific response
- a specific response.

Figure 8.2
An overview of the body's defences against pathogens

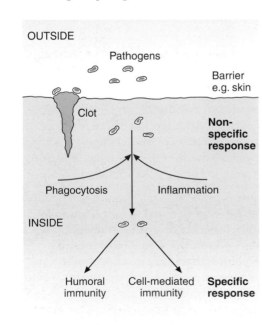

Barriers to prevent entry

Skin, the outermost layer of the body, forms a natural barrier and prevents the entry of microorganisms. The surface of the skin is dry and composed of dead cells, which contain the protein **keratin**. This cannot easily be digested and so bacteria, viruses and fungi cannot penetrate it. The outermost cells are also continuously flaking off, taking the bacteria with them. An oily substance called **sebum** produced by the skin contains substances, such as fatty acids, which lower the pH of the skin and inhibit the growth of microorganisms. The sebum, however, does not harm the non-pathogenic microorganisms which live on our skin without causing any damage. These helpful microorganisms play a role in preventing disease by competing successfully for positions on the skin. In some cases, they produce substances that inhibit the growth of other microorganisms. **Sweat** produced by sweat glands contains the enzyme **lysozyme**, also found in **tears**, which digests the cell walls of bacteria. The only way in which these pathogens can enter the body through the skin is when the surface is broken.

The **gut** also has defensive mechanisms. The saliva in the mouth contains the enzyme lysozyme. Hydrochloric acid in the stomach destroys many ingested bacteria.

In the **gas exchange** mucus is secreted. This is a sticky substance that traps microorganisms, which are constantly swept upwards by cilia to pass out of the body or to be swallowed into the stomach.

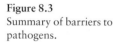

Figure 8.3
Summary of barriers to pathogens.

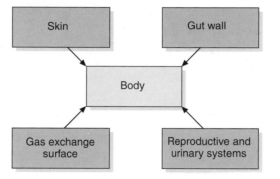

Pathogens can also enter the body through the lining of the **urinary** and **reproductive tracts**. Each of these has its own protective mechanism.

Despite these defences some pathogens still gain entry. Once inside the body, they are much more difficult to deal with: not only must the body distinguish them from its own cells, but it must also destroy them without causing damage to its own tissues.

Q 2 Smoking leads to paralysis of the cilia, which line the airways. Suggest why smokers are more likely to get lung infections than non-smokers.

Non-specific response

The body responds immediately to injury and the presence of pathogens in a number of ways. These include:

- producing an inflammatory response
- producing phagocytes
- causing the blood to clot.

Inflammatory response

Cells of the immune system are widely dispersed throughout the body, but when there is a local infection, they accumulate in that area. **Histamine** is released into the area of a wound by a type of white cell. This has a number of effects, including making the nerve endings more sensitive, which causes the area to become painful to touch.

Other effects increase the number of white cells in that area. These include:

- vasodilation, which increases the local blood flow, causing the area to become red
- increased permeability of blood capillaries, causing an increase in tissue fluid formation and swelling.

145

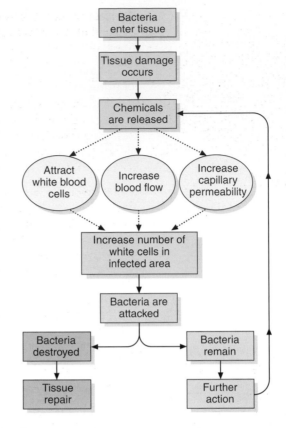

Figure 8.4
Summary of the inflammatory response.

Phagocytosis

Certain white cells are capable of engulfing bacteria by a process known as phagocytosis and contain digestive enzymes within **lysosomes**. They also contain a variety of substances such as hydrogen peroxide which kill bacteria.

During the early part of an infection, these white cells collect at the site of infection. They are attracted by substances released by bacteria and by host cells. The white cells stick to the bacteria and kill them. The bacterial cells are digested by enzymes. The products of digestion are absorbed into the cytoplasm of the white cells.

The white cells have a short life and after killing and digesting a few bacteria, they die. Dead white cells often collect at a site of infection to form **pus**.

Q 3 Histamine increases the permeability of blood capillaries. How will this assist phagocytosis?

Blood clotting

If the skin is damaged, the first response of the body is to seal the wound through the mechanism of blood clotting. In theory even a minor cut could endanger life because of blood loss. In normal circumstances this does not happen because the body has a highly efficient blood clotting system. This process also seals the damaged vessels and prevents the entry of pathogens into the blood.

The process of forming a blood clot follows a complex sequence of events. Important components of the blood-clotting system are the clotting factors, which include proteins and enzymes, and the circulating platelets.

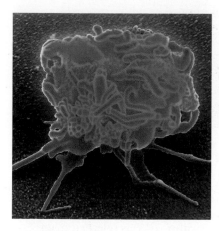

Figure 8.5
An activated platelet.

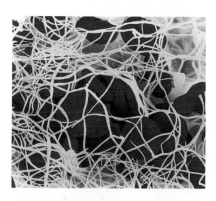

Figure 8.6
Scanning electronmicrograph of a blood clot. The fibrin strands have trapped the red blood cells in a mesh.

Platelets are not strictly cells. They are actually fragments of a cell, pinched off from large cells that sit permanently in the bone marrow. These cells send out millions of tiny fragments. Platelets, which are shaped like flattened discs, do not have a nucleus and so cannot reproduce themselves.

Platelets only exist in the circulation for 5–7 days before they are removed. During this time they circulate in the blood vessels continuously, reacting to any signs of damage. Tissue damage can take many forms, but if a blood clot is to develop, the blood has to be exposed to tissue components such as collagen or muscle.

Something quite remarkable happens to platelets when they detect damage. Within seconds they change from smooth flattened discs to tiny spheres with many long thin arms like the tentacles of an octopus. Their surface becomes sticky and they adhere to cut surfaces of tissue, especially the area immediately surrounding broken blood vessels. In this state they are said to be **activated** (See Figure 8.5).

When a platelet is activated, it releases a number of substances to initiate a rapid clotting process and thus reduce the amount of blood lost. These include

- **serotonin,** which causes the smooth muscle of the arterioles to contract and so narrows the vessel, cutting off the blood flow to the damaged area

- **thromboplastin,** an enzyme that sets in progress the series of events that leads to the formation of a clot.

In the blood, if there are enough **calcium ions**, the thromboplastin catalyses the conversion of an inactive enzyme found in the blood plasma, **prothrombin** into its active form, **thrombin**:

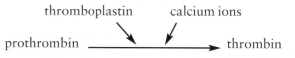

The thrombin acts on another plasma protein. This is called fibrinogen and it is converted to fibrin:

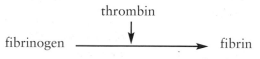

The fibrin is insoluble and consists of long strands. These strands form a web over the damaged area and trap further platelets and blood cells, forming a clot.

One remarkable feature of blood clot formation is the speed at which it occurs. From start to finish it takes about 20 seconds.

Clotting is an example of a 'cascade' effect. The release of one molecule of thromboplastin can catalyse the conversion of thousands of molecules of prothrombin to thrombin. In turn, one molecule of thrombin can produce the conversion of hundreds of molecules of fibrin. The whole process is summarised in Figure 8.7.

Figure 8.7
The steps in blood clotting.

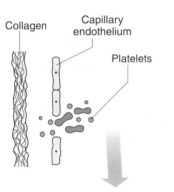

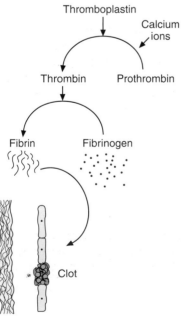

If pathogenic viruses, bacteria or other microorganisms penetrate these defences, they enter the bloodstream and here they encounter the body's immune system.

Specific response

This system consists mainly of white cells circulating in the blood and lymph, together with those contained in the spleen, liver and lymph nodes. It gives long-lasting immunity against anything the body recognises as foreign.

Before we look at the process in detail, we need to look at some of the terms we use in describing the immune system and the way it works.

Antigen

This is a molecule that triggers an immune response. Although antigens are usually proteins, other types of molecule such as polysaccharides, lipids and even nucleic acids can act as antigens. A **self-antigen** is a molecule found on the surface of your own cells to which you do not respond. However, it will cause a response if introduced into another person. It has a specific shape and is present on every cell. It helps with self-recognition. A **non-self-antigen** is a molecule found on cells entering your body that are not yours, e.g. bacteria, viruses or even another person's cells. It will produce an immune response in your body.

Antibody

An antibody acts against foreign bodies. It is a molecule secreted by certain B-lymphocytes in response to stimulation by the appropriate antigen. Antibodies are types of proteins known as immunoglobulins. They have a unique antigen-binding site.

The body's specific response is based on its ability to recognise and respond to non-self-antigens. The response depends on the activity of various types of white blood cells. The main types of cells involved in defence are:

Figure 8.8
Antigens present on cells within your own body and those present on foreign cells.

- phagocytes, e.g. macrophages
- lymphocytes.

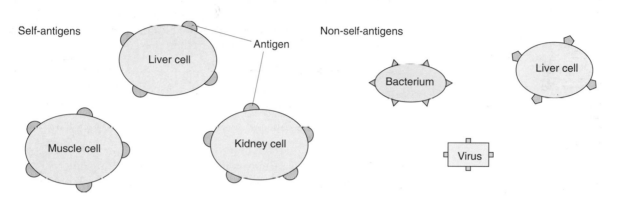

Phagocytes

Phagocytes are continually produced by bone marrow throughout life. They are stored there but can leave the blood to be distributed around the body. Macrophages circulate in the blood and pass into organs such as the lungs, kidney and lymph nodes, where they tend to remain. They are large long-lived cells that remove foreign matter from organs. They play a crucial role in initiating the specific immune response.

Lymphocytes

There are two groups of lymphocytes:

- B-lymphocytes (called B-cells)
- T-lymphocytes (called T-cells).

Both types of cell must go through a maturing process before they can take part in the specific immune response. This takes place before birth and results in the production of many different types of each group of lymphocyte.

B-lymphocytes

Each B-cell is genetically programmed to develop just one type of specific receptor on its cell surface membrane. These receptors have the same shape as the antibody that will be produced by the B-cell. The antibody molecules are able to combine specifically with antigens.

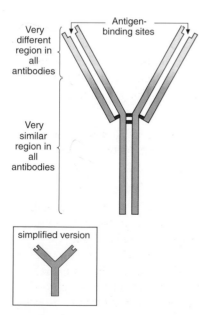

Figure 8.9
The structure of an antibody, showing its unique antigen-binding site.

149

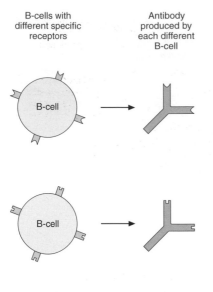

Figure 8.10
B-lymphocytes are present in a variety of forms, each having the ability to produce a specific antibody related to the receptors on its surface.

During the maturation process B-cells develop to give as many as 10 million different possible variants. This gives the specific immune system the ability to respond to any type of pathogen that enters the body.

B-cells are unable to respond to the presence of an antigen on their own: they rely on a type of T-lymphocyte known as a T-helper cell.

T-lymphocytes

T-cells develop specific surface receptors called T-cell receptors. A very important sub-group of T-cells are T-helper cells, which have an active role to play in the specific immune response.

Humoral immunity

This is an immune response to a specific pathogen, such as a bacterium. When bacteria enter the bloodstream they circulate around the body until they eventually reach the lymph nodes. Here they are ingested by the macrophages. The macrophages cut out the surface molecules of bacterial cell walls and membranes. These pieces are the antigens of the bacteria and the macrophages display them on their cell surface membranes. This is called **antigen presentation** (see Figure 8.11).

Inside the lymph nodes the macrophage joins with specific T-helper cells and B-cells that have receptors in their membrane complementary in shape to the antigen presented by the macrophage. During antigen presentation (see Figure 8.12(a)) the macrophage selects the T-helper cells and B-cells that have membrane receptors that are complementary in shape to the antigens it has exposed. This is known as **clonal selection**. The selected T-helper cells are induced to secrete cytokines, which activate the selected B-cells to divide to form a number of identical cells called **clones** (see Figure 8.12(b)).

Clonal expansion produces large numbers of identical B-lymphocytes with the ability to destroy this particular pathogen. Bacteria divide so rapidly in the ideal conditions found within the host's body that unless huge numbers of B-lymphocytes are produced, the bacteria could cause damage.

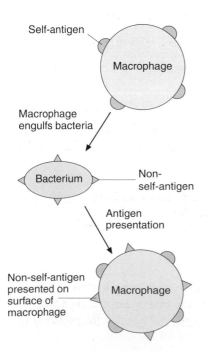

Figure 8.11
When a non-self-antigen on a bacterium is present, the macrophage will engulf the bacterium and present the non-self-antigen on its surface.

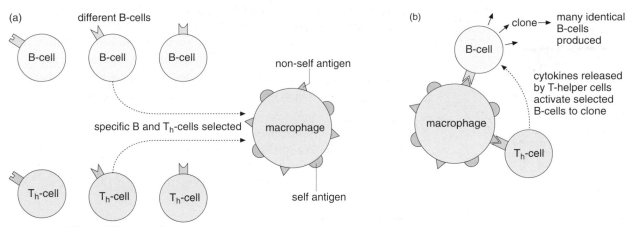

Figure 8.12
(a) Specific B-lymphocytes and T-helper cells are selected, depending on their specific surface receptors.
(b) When selected, the T-helper cells secrete a chemical that stimulates the B-cells to clone.

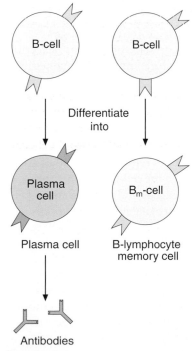

Figure 8.13
Once stimulated to clone, the B-cells differentiate into plasma cells and B-lymphocyte memory cells.

Some of the clones of activated B-cells will differentiate into **B-lymphocyte memory cells** and some into **plasma cells** (see Figure 8.13). The B-lymphocyte memory cells are stored in the lymphatic tissue, while the plasma cells very quickly produce a specific antibody at a rate of up to several thousand a second. The antibodies produced circulate in the blood and lymph or secrete antibodies onto the surfaces of mucous membranes, such as those found lining the lungs. These specifically shaped antibodies bind to the antigen on the surface of the invading bacteria.

Different antibodies work in different ways:

* **agglutination** makes the pathogens clump together

* **antitoxins** neutralise the toxins produced by the bacteria

* **lysis** digests the bacterial membrane, killing the bacterium

* **opsonisation** coats the pathogen in protein that identifies them as foreign cells.

Plasma cells are short lived. After a few weeks their numbers decrease but the antibody molecules they have secreted remain in the blood for some time, maintaining immunity. Eventually, the concentration of antibodies will also decrease.

Q 5 How would injection of serum containing antibodies help someone suffering from meningitis?

Immunological memory

When a bacterial infection occurs and an antigen is presented for the first time, clonal selection and clonal expansion take time to occur. Once B-cells have differentiated into plasma cells, the specific antibodies that

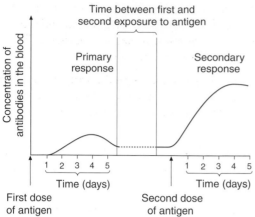

Figure 8.14
The primary and secondary responses to the same antigen.

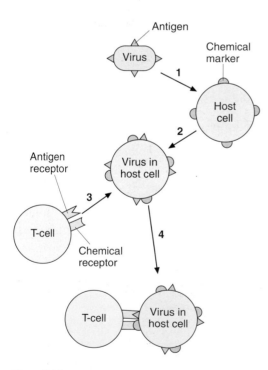

Figure 8.15
T-lymphocytes have to recognise both the specific antigen and a chemical marker to become sensitised.

they secrete can be detected in the blood. This **primary response** lasts several days or weeks and then the concentration of antibody decreases as the plasma cells stop secreting them. There is little point in producing antibodies to a pathogen that is not present, so as the infection subsides, plasma cells die but memory B-cells are left in the body.

If another infection of the same pathogen should occur, the same antigen is reintroduced. There is now a more rapid response, producing a higher concentration of antibody. This **secondary response** occurs quickly because of the presence of memory B-cells formed during clonal expansion in the primary response, which have remained in the lymphatic system.

When the same antigen is presented by a macrophage, the relevant memory B-cells divide rapidly to produce plasma cells. This process is repeated whenever the same antigen is identified. The secondary response is much faster because there are many memory B-cells, which can produce many plasma cells and the appropriate antibody. These destroy the pathogen before it has the chance to cause any symptoms to occur.

Memory cells are the basis of immunological memory; they last for many years, often a lifetime.

It is possible to suffer repeated infections from a single pathogen because pathogens sometimes occur in different forms, each having minor changes in the shape of the antigen, due to a possible mutation, and therefore requiring a primary response.

Each time we have an infection with a different pathogen, bearing different antigens, antigen presentation, clonal selection and clonal expansion must occur before immunity is gained.

Cell-mediated response

As seen earlier, the humoral response deals mainly with bacteria, which do not enter our cells. However, other pathogens can enter our cells and are much more difficult to remove. Antibodies are of limited use because they cannot cross the cell surface membranes to reach the pathogen inside. T-killer cells and macrophages can only remove these pathogens by destroying infected host cells. This process is known as **cell-mediated immunity**, in which the cells of the immune system directly kill pathogens, unlike the case of **humoral immunity** in which antibodies work at a distance from the plasma cells that secrete them. T-lymphocytes are involved in this response. Each T-lymphocyte cell expresses a specific surface receptor called a T-cell receptor. These receptors can only recognise antigens that are present on the surface of cells. There are two main types of T-lymphocytes:

- T-helper lymphocytes
- T-killer lymphocytes.

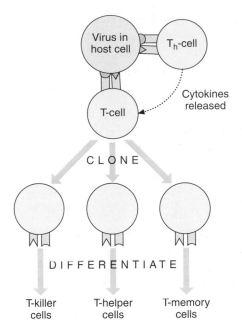

T-lymphocytes, like B-lymphocytes, have to make contact with their matching antigen before they can be stimulated and the T-lymphocytes have specific receptor molecules on their cell surface membranes to enable them to do this.

T-lymphocytes can only recognise their specific antigen if it is on the surface of a human cell next to a **chemical marker**. These chemical markers are examples of the major histocompatibility complex proteins.

Once the T-lymphocytes have attached to their specific antigen, they become sensitised and divide rapidly to form a clone. These clones differentiate into a number of different types of T-cell.

The T-killer cells recognise the specific antigen–chemical marker complex formed only on body cells infected by intracellular pathogens, e.g. viruses. They destroy these cells and others, such as those within transplanted organs.

Figure 8.16
A sensitised T-cell clones and differentiates into different types of T-cell.

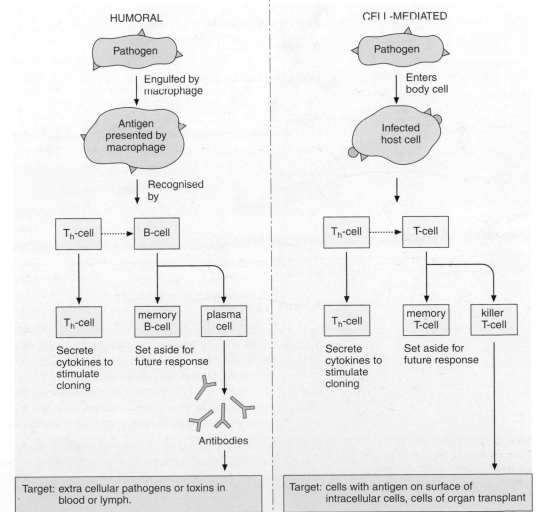

Figure 8.17
Difference between humoral and cell-mediated immune response.

T-killer cells can secrete **perforins**, which form holes in the target cell membrane. As a result the target cell bursts or may secrete **toxins** that disrupt organelles and damage DNA.

Immunisation

Immunisation refers to various processes that make use of the immune system and increase protection against specific diseases. Although drugs can be used to treat microbial infections once they are present in the body, an alternative approach to breaking the cycle of disease is to prevent it from developing in the first place. This is what immunisation does.

When we are exposed to a pathogen and suffer from a disease, we develop a natural immunity to that pathogen. Having dealt with the pathogen once, our immune system can destroy it before it causes disease following a subsequent infection. Vaccination enables us to develop this immunity artificially by exposing our immune system to the pathogen in a safe way. Antibodies and memory cells can be produced without us having the disease.

There are two main types of immunity:

- passive immunity
- active immunity.

Passive immunity

Passive immunity can be gained while a fetus is developing inside the uterus of its mother . Antibodies pass from mother to fetus across the placenta. Passive immunity can also be gained from the milk when breast-feeding. The immunity given in this way is short lived.

Artificial passive immunity occurs when you are supplied with antibodies, rather than producing your own.

As antibodies are proteins and proteins are continually being broken down by the liver, they only last for a few weeks. Thus the disadvantage of passive immunity is that the protection, although immediate, is relatively short lived, lasting only as long as the antibodies remain in the blood.

Active immunity

Active immunity is any method in which the body is stimulated to produce its own immune response. This can only be achieved by introducing an antigen into the body. Active immunity can be established by immunisation. A small quantity of antigen in the form of a **vaccine** is introduced into the body. This activates the B-lymphocytes and memory cells are produced. These memory cells are ready to spring into action if the person is exposed to the disease-causing organism on a later occasion.

Before injection the antigens are treated in a way that maintains their shape but makes them relatively harmless. Such treated antigens are called vaccines. Five different kinds of vaccine are widely used.

Dead pathogens

Dead microorganisms, which still have the same antigens on their surfaces but cannot obviously reproduce, are injected into the body. The antigens cause the production of specific antibodies, but only produce a few memory cells. Another dose of the original vaccine given some time after the original dose, known as a **booster**, will therefore be required. Examples of this type of vaccine are those for whooping cough and cholera.

Live pathogens

Live pathogens, which have been weakened so they no longer cause disease, are known as **attenuated** pathogens. As these pathogens are still living, they are able to reproduce and therefore stimulate a full immune response. This usually provides long-lasting immunity. Examples of this type of vaccine are rubella and tuberculosis vaccines.

Chemically modified toxins (toxoids)

Toxoids are mild forms of the toxins produced by some bacteria. They are not poisonous, yet they contain antigens, which can stimulate the production of antibodies. Toxoids form the basis of the tetanus and diphtheria vaccines.

Extracted antigens

In modern influenza vaccines, antigens from the viral coat are separated and inoculated. There is no risk of developing the disease, but an effective immune response is stimulated.

Genetically engineered antigens

This method involves inserting the genes for the antigens of the pathogen into a relatively harmless microorganism. These microorganisms produce the appropriate antigen without causing the disease. The vaccine against hepatitis B is produced in this way.

Extension box 2

In many millions of people, immune responses are made against substances which are normally harmless. Substances that provoke these abnormal responses are known as **allergens**, and the response itself is called an **allergy**. Common allergens are pollen, a variety of drugs and foods, dust mites and cosmetics. Some responses to allergens start within minutes, others may be delayed. Whatever the time-scale, allergens provoke an inflammation response.

Some people are genetically inclined to allergies. Besides this, infections, emotional stress or changes in temperature may trigger reactions that otherwise might not occur.

Q 6 What do you think it means when products such as make-up are labelled hypoallergenic?

On an exposure to an allergen, special types of antibodies are produced. These bind to cells called mast cells. The mast cells secrete a number of substances as a result and it is these substances which trigger the inflammatory response. They also stimulate the secretion of mucus and cause airways to constrict. In asthma and hay fever, congestion, sneezing, a runny nose and laboured breathing are symptoms of the allergic response. One of the substances produced is histamine so antihistamines are often used to relieve the short-term symptoms of allergies.

Figure 8.18
Hay fever – a demonstration of one of the effects of an allergic response

In some cases, inflammatory reactions proceed throughout the body and trigger a life-threatening condition called **anaphylactic shock**. For example, a person who is allergic to bee venom can die within minutes of a single sting. Air passages leading to the lungs constrict. Fluid escapes too from dilated capillaries made leaky by the release of large amounts of histamine. Blood pressure drops sharply, which may lead to circulatory failure. These symptoms happen very quickly and can be fatal unless immediate treatment with adrenaline and antihistamines is available. The adrenaline rapidly constricts the blood vessels, thereby dramatically reducing the leakage of blood into the tissue. Desensitisation programmes may help over the long term.

Summary

This chapter should help you to be able to:

- describe phagocytosis and the subsequent destruction of ingested pathogens

- describe the roles of thromboplastins, prothrombin, plasma enzymes, calcium ions and fibrinogen in blood clotting

- give definitions of an antigen and an antibody

- describe the immunological response of B-lymphocytes and T-lymphocytes when stimulated by the appropriate antigen

- describe the essential differences between humoral and cellular (cell-mediated) responses

- describe the role of plasma cells and memory cells in producing primary and secondary responses

- describe passive immunity, both natural and acquired

- describe vaccination and immunisation

- describe different methods of forming vaccines from attenuated and dead microorganisms, and by genetic engineering.

Assignment

The beginning of the end

Smallpox was a dreadful disease. The virus that caused it spread rapidly from one person to another and epidemics were frequent. The early symptoms of fever were followed by the appearance of pustules or sores all over the body. Smallpox was often fatal. Even if a person recovered, his or her face was often left permanently disfigured by pits and scars where the pustules had been (Figure 8.19).

Smallpox has now been eradicated, the result of a world-wide programme of vaccination. In this assignment we will look at the work of one of the pioneers of modern medicine – Edward Jenner. It was thanks to his efforts that an effective vaccine against smallpox was developed. In 1801, Jenner published a scientific paper with the title, "The origin of vaccine inoculation". We will look at some extracts from this paper. Remember that it was written 200 years ago so the language is a little different from that we use today. The first extract tells how he became interested in the relationship between smallpox and another disease, cow pox.

My inquiry into the nature of Cow Pox commenced upwards of twenty-five years ago. My attention to this singular disease was first excited by observing that among those in the country I was frequently called upon to inoculate, many resisted every attempt to give them Small Pox. These patients I found had undergone a disease they called the Cow Pox.

Figure 8.19
A smallpox victim. You can see the pustules on this child's body. Even if she recovers, she will be permanently scarred.

1. At the time Jenner made these observations, some people tried to protect themselves against smallpox by getting inoculated with material from a smallpox pustule. This was a rather dangerous thing to do.

   (a) Use your knowledge of B-lymphocytes and memory cells to explain how inoculation with material from a smallpox pustule could protect against a later attack of smallpox.

   (4 marks)

   (b) Suggest why this was a rather dangerous thing to do.

   (1 mark)

2. Explain why those "who had undergone a disease they called the Cow Pox" "resisted every attempt to give them Small Pox".

   (2 marks)

Jenner encountered a number of set-backs in his investigation. The following passage describes one of them.

In the course of the investigation of this subject… I found that some of those who seemed to have undergone the Cow Pox nevertheless, on inoculation with the Small Pox, felt its influence just the same as if no disease had been communicated to them by the cow… This for a while damped but did not extinguish my ardour; for as I proceeded, I had the satisfaction to learn that the cow was subject to some varieties of spontaneous eruptions on her teats; that they are all capable of communicating sores to the hands of the milkers; and that whatever sore was derived from the animal was called in the dairy the Cow Pox. Thus I . . . was led to form a distinction between these diseases.

3. Explain in your own words:

   (a) what led to Jenner claiming that his ardour was "for a while damped";

   (b) how Jenner explained the problem that he encountered.

Now read the next extract. It describes the first experiment that Jenner carried out with material from a cowpox pustule.

The first experiment was made upon a lad of the name of Phipps, in whose arm a little vaccine virus was inserted, taken from the hand of a young woman who had been accidentally infected by a cow (Figure 8.20). Not withstanding the resemblance which the pustule, thus excited on the boy's arm, bore to variolus inoculation (i.e. using material from a smallpox pustule) yet as the indisposition attending it was barely perceptible, I could scarcely persuade myself the patient was secure from the Small Pox. However, on his being inoculated some months afterwards, it was proved he was secure.

DNA and inheritance

The sequence of bases in DNA controls the order of amino acids in proteins that are made by a cell's ribosomes. Without the correct sequence of bases, a cell will not be able to make a particular protein. This is important when we remember that many of the proteins made in a cell are enzymes. These enzymes control the chemical reactions that occur in a cell. Reactions in cells are referred to as **cell metabolism**. The sequence below shows how several enzymes can control a series of chemical reactions. Such a sequence is called a **metabolic pathway**.

$$\text{enzyme 1} \qquad \text{enzyme 2} \qquad \text{enzyme 3}$$
$$\text{substance A} \rightarrow \text{substance B} \rightarrow \text{substance C} \rightarrow \text{substance D}$$

The length of DNA that carries the code for a particular protein is called a **gene**. If the gene for a particular protein is present in a cell's DNA, the cell will be able to make that protein. If the protein is an enzyme, it will have an effect on the cell's metabolism. Often the effect of an enzyme on an organism can be observed or detected chemically. Properties of an organism that can be detected are called its **phenotype**. Hair colour is part of the phenotype that is easily detected in humans. Blood group is part of the phenotype that is less easy to detect than hair colour. Like all aspects of our phenotype, these features are influenced by metabolic pathways. Specific enzymes control these pathways. We can now see the link between DNA and phenotype (Figure 9.9).

Figure 9.9
By controlling the production of enzymes, DNA affects an organism's phenotype.

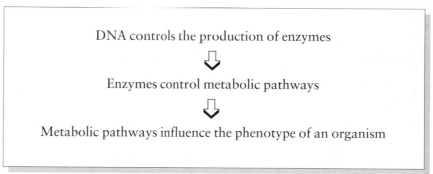

Note that the environment also affects phenotype. This is easy to imagine if you think of hair colour. Although inherited, many people use bleaches or dyes to change their hair colour. You can become blonde even if your genetic code is for brown hair!

We will now look at the effect that two 'defective' genes may have on our phenotype. The 'defective' genes result in two common human disorders: cystic fibrosis, which affects 1 in 2500 people in the UK, and phenylketonuria, which affects 1 in 10 000 people in the UK.

key term

A **gene** is a length of DNA that carries the genetic code for a single protein. Different forms of the same gene are called **alleles**. (For example, brown hair and blonde hair are controlled by alleles of the gene for hair colour.)

Cystic fibrosis

Cystic fibrosis affects glands that produce mucus. The glands of affected people produce mucus that is much thicker than usual. In the lungs, this thick mucus blocks airways and increases the likelihood of lung infections. In the pancreas, the mucus blocks ducts that normally let digestive enzymes pass to the gut. The pancreas reacts by forming fluid-filled sacs, called cysts. Eventually the structure of the pancreas degenerates and becomes fibrous. It is these cysts and fibres that give the disorder its name. The liver is affected in a similar way to the pancreas.

Q 7 **Suggest why people with cystic fibrosis have an increased likelihood of lung infections.**

Cystic fibrosis is caused by a defective gene. The normal form of the gene codes for a protein called the **cystic fibrosis transmembrane regulator** (CFTR). CFTR is found in the surface membranes of cells in the lungs and the gut. CFTR is an ion channel (see Chapter 2). It allows the transport of chloride ions out of cells lining the lungs, pancreas and liver. By osmosis, water follows the chloride ions and thins the mucus produced by these cells. Biologists have found over 250 defective forms (alleles) of the CFTR gene. The most common (accounting for 75% of sufferers) results in the CFTR protein being made with one crucial amino acid missing. That amino acid is phenylalanine. Figure 9.10 summarises the effects of the normal CFTR allele and this mutant allele.

Figure 9.10
A flow diagram to show the effects of the normal and the mutant alleles.

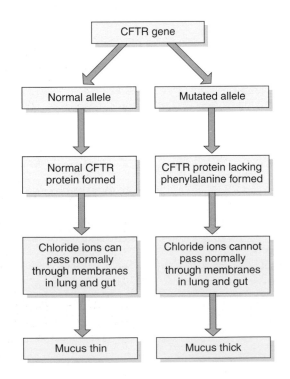

Phenylketonuria

Phenylketonuria (PKU) also involves the amino acid phenylalanine. Normally, phenylalanine is converted to tyrosine, another amino acid. The conversion is controlled by an enzyme called phenylalanine hydroxylase.

phenylalanine hydroxylase

phenylalanine $\longrightarrow$ tyrosine

A PKU sufferer (**phenylketonuric**) is unable to convert phenylalanine into tyrosine. Instead, phenylalanine accumulates in the body. Although not harmful to adults, phenylalanine is harmful to the developing brains of young children. An infant with PKU could suffer severe mental retardation. For this reason, all babies in the UK are screened for PKU. This involves testing a drop of blood taken from the baby's heel (the Guthrie test).

Phenylketonurics are unable to produce the enzyme phenylalanine hydroxylase. Their genes for phenylalanine hydroxylase carry the wrong code. Although a protein is made when this code is transcribed and translated, it does not have the necessary enzyme activity.

Q 8 Cans of diet cola carry the following warning label: 'Phenylketonurics : Contains phenylalanine.' Suggest the reason for this warning label.

Summary

- Nucleic acids carry the genetic code that cells use to make proteins. The code determines the sequence of amino acids in a protein.

- Nucleic acids are long chains of smaller units called nucleotides. Each nucleotide has a five-carbon sugar, a phosphate group and an organic base.

- Deoxyribonucleic acid (DNA) is made of two chains of nucleotides: all its nucleotides carry the five-carbon sugar deoxyribose. Ribonucleic acid (RNA) is made from a single chain of nucleotides: all its nucleotides carry the five-carbon sugar ribose.

- There are two groups of organic bases: purines and pyrimidines. Adenine and guanine are purine bases whereas cytosine, thymine and uracil are pyrimidine bases. Adenine, cytosine, guanine and thymine are found in DNA. Adenine, cytosine, guanine and uracil are found in RNA.

- Provided they have the complementary base sequence, strands of nucleic acid can be held together by hydrogen bonds between bases. These base pairs are highly specific. In DNA adenine always pairs with thymine and cytosine always pairs with guanine. RNA forms similar base pairs, except that adenine pairs with uracil.

- Hydrogen bonds between base pairs hold together the two complementary strands in a DNA molecule. They also hold parts of RNA molecules together where the single strand has folded back on itself.

- Protein synthesis can be considered in two stages: transcription and translation. During transcription, the base sequence of part of a DNA molecule is copied by a molecule of messenger RNA (mRNA). Each group of three bases on the mRNA molecule, called a codon, codes for a particular amino acid. This is the genetic code.

- Introns are lengths of nucleic acid that carry a nonsense code. Although copied from DNA to mRNA during transcription, these introns are removed from the mRNA before it is used to make proteins.

- During translation, the base sequence of mRNA is 'read' by ribosomes and used to assemble a sequence of amino acids. Transfer RNA (tRNA) molecules are involved in this process. They have an anticodon, complementary to each codon, that enables them to carry amino acids to the ribosomes.

- Specific 'start' codes ensure that the genetic code is non-overlapping. This means that it can only be read in a unique sequence of three bases at a time.

- DNA is copied so that it can be passed from cell to cell during cell division. During semi-conservative replication, the two strands in the parent DNA break away from each other. By base pairing, free nucleotides assemble along the exposed single DNA strands. Because the base pairing is specific, each new strand is an exact copy of the old partner strand. DNA polymerase joins together the free nucleotides.

- The length of DNA that carries a code for one protein is called a gene.

- Because many proteins are enzymes, the DNA determines the metabolic reactions that occur in cells. In this way, DNA contributes to the phenotype of organisms.

- Spontaneous changes, called mutations, sometimes occur in DNA. These mutations often disrupt metabolic pathways. Cystic fibrosis and phenylketonuria are two examples of disorders caused by defective genes.

Assignment

Although our knowledge of biology has grown enormously over the last 50 years, each new discovery only adds a little to our overall understanding. In order to interpret our discoveries and gain a better understanding, we often construct models. These models can then be used to make predictions that can be tested further. You will have already come across models in your biology course. One of the simplest is the 'lock and key' model used to explain the way in which enzymes work. Testing this model suggested improvements that could be made and as a result we now think that induced fit provides a better explanation of the way in which enzymes work.

Our knowledge of the structure of DNA is the result of the work of many biologists including James Watson and Francis Crick working in Cambridge. They reviewed a lot of information and used it to produce a molecular model of DNA which they used to test their predictions. In this assignment you will also produce a model of DNA. Although it will be constructed in a simple way, it should enable you to make predictions about its structure and properties. You can make the model and answer Questions 1 to 4 before you start to study the topics of DNA and protein synthesis but you may wish to leave Question 5 until you have found out rather more about this important molecule.

Figure 9.11
The four nucleotides which form a DNA molecule. Each nucleotide consists of a five-carbon sugar, deoxyribose, a phosphate group and a base. A different base is found in each of these nucleotides.

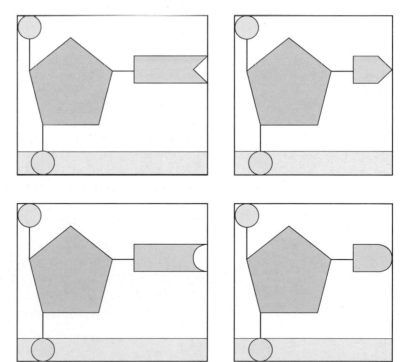

- Figure 9.11 shows the four basic units or nucleotides which form a DNA molecule. Before you start making your model, you will need about 10 copies of each of these nucleotides. The simplest approach is probably to cut them out from photocopies.

Figure 9.12
Molecular structure of the four different bases found in a DNA molecule.

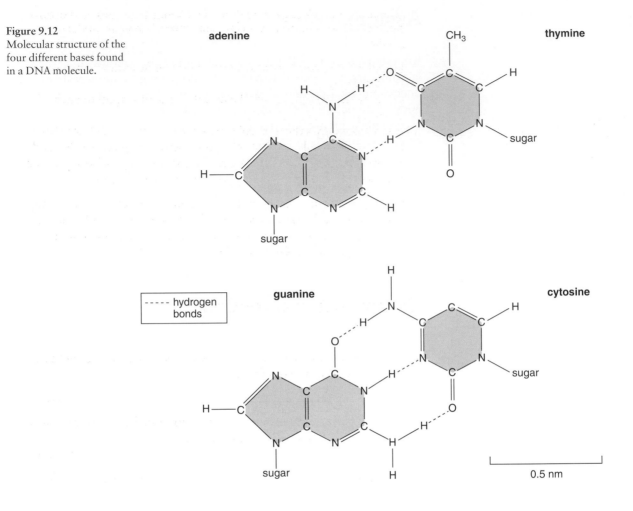

- Now produce a single polynucleotide chain. Mix your nucleotides together and pick out 20 at random. Stick these together so that the lower phosphate group of one overlaps the upper phosphate group of the next. This means that the top of the second nucleotide should completely overlap the shaded area at the bottom of the first nucleotide.

1 Explain why this polynucleotide chain can be described as a polymer.

2 (a) The first nucleotide that you picked out could have been any one of the four different types. If you selected two nucleotides at random from your initial collection, how many different combinations would it have been possible to have selected? Remember, it is possible that the two nucleotides might be the same.

   (b) Use your answer to Question 2 (a) and your knowledge of protein structure to explain how DNA is suited to its role of coding for different proteins.

3 Watson and Crick suggested that a DNA molecule consisted of two polynucleotide chains parallel to each other. Use your model to predict:

(a) which nucleotide bases in one chain will fit with which bases in the other;

(b) how the two chains will be orientated with respect to each other.

4 Figure 9.12 shows the molecular structures of the four different bases found in a molecule of DNA. Use your answer to Questions 3(a) and Figure 9.12 to explain why the bases in the two polynucleotide chains of a molecule of DNA can only combine in a specific way.

5 It is important to remember that what you have constructed is only a model and it differs considerably from a real DNA molecule. In what ways does your model differ from our present ideas about the structure of a DNA molecule?

Examination questions

1 (a) (i) State how the individual nucleotides are held together in a single strand of DNA.

(1 mark)

(ii) Name the type of bond that holds together the two strands of nucleotides in a DNA molecule.

(1 mark)

(b) Table 9.5 shows the sequence of bases on part of a molecule of mRNA.

Base sequence on coding strand of DNA									
Base sequence on mRNA	A	G	C	U	G	U	A	C	U

Table 9.5

Complete the table to show the base sequence of the coding strand of a molecule of DNA.

(1 mark)

(c) (i) A particular strand of mRNA is 450 bases long. How many amino acids would be present in the protein coded by this mRNA?

(1 mark)

(ii) Explain your answer.

(2 marks)

2 Table 9.6 shows the proportion in bases in the nucleic acid of two animals.

Nucleic acid	Adenine	Cytosine	Guanine	Thymine
DNA in gene of animal A	26%			
mRNA resulting from gene in animal A	29%	22%	18%	0%

Table 9.6

(a) (i) Complete the table to show the percentage of bases in the DNA of this animal.

(1 mark)

(ii) Explain how your knowledge of DNA structure enabled you to calculate your answer.

(2 marks)

(b) The percentage composition of mRNA is not the same as the DNA gene from which it is made. Use your knowledge of mRNA formation to explain:

(i) the result for thymine;

(2 marks)

(ii) the result for adenine.

(1 mark)

3 (a) Draw a diagram to show the structure of a single mRNA nucleotide.

(2 marks)

(b) Describe how mRNA contains a code for protein structure.

(4 marks)

10 The Cell Cycle

In 1951, scientists at the Johns Hopkins University in the USA were trying to grow human cells outside the body. Such cells would enable them to study cell physiology. The scientists would also be able to study diseases without having to experiment directly on humans. At first, all the cell cultures they started died within a few weeks.

Finally, one cell culture proved successful. These cells were code named HeLa, from the first two letters of the donor's first and second name, Henrietta Lacks. The reason that HeLa cells grew so well in the laboratory was that they were cancer cells. As the HeLa cells continued to divide rapidly in the laboratory, so they did within Henrietta's body. Six months after the first diagnosis, cancer cells had spread through Henrietta's body. Two months later, at the age of only 31, she was dead.

Although Henrietta had died, some of her cells lived on in the laboratory at Johns Hopkins University. Subcultures of these cells were sent to research biologists in other laboratories who, in turn, passed them on to others. In a few years, HeLa cells were cultured in research laboratories throughout the world. Some were even sent into space aboard the *Discoverer XVII* satellite.

Every year, research that is published in hundreds of scientific papers is based on work with HeLa cells. The uncontrollable cell division that killed Henrietta some 50 years ago continues unabated in laboratory cultures around the world.

Cell division

We saw in Chapter 1 that all organisms are made of cells. New cells can only be made when existing cells divide. Taken together, these two ideas form the cell theory. Although all cells have the potential to divide, many cells in a eukaryotic organism lose this ability. For example:

- cell division replaces the entire lining of your small intestine every five days

- your liver cells divide only to repair damage, and then stop dividing

- your nerve cells do not divide.

Humans are eukaryotic organisms, i.e. their cells have a nucleus. There are two types of cell division in eukaryotic cells: mitosis and meiosis. In this chapter you will learn about only one of them: mitosis. Cell division in prokaryotic cells does not involve either mitosis or meiosis.

key term

The **cell theory** states that all organisms are made of cells and that new cells can only arise by the division of existing cells.

Mitosis

During mitosis, a parent cell divides to produce two daughter cells. Each daughter cell contains some of the cytoplasm of the parent cell, including its organelles. The most important feature of mitosis is that the daughter cells each contain the same genetic code as the parent cell. In other words, mitosis produces cells that are genetically identical to each other and to the parent cell. The parent cell can do this because it has copied its own chromosomes prior to cell division.

Q 1 Name the process by which a parent cell copies its chromosomes prior to mitosis.

Mitosis occurs rapidly during growth, when new cells are made. The growth of a fertilised egg to produce a fetus and the growth of a child into an adult are the result of mitosis. Mitosis also occurs rapidly in many types of mature human tissue. These include epithelial tissues, such as the outer layer of the skin (epidermis) and the lining of the gut. The new cells produced in skin replace those that have been lost, for example, by friction. If the skin has been cut, mitosis produces new cells that repair the damage caused by the cut. However, not all mature human cells can divide by mitosis. Mature nerve cells are one example of human cells that have lost the ability to divide by mitosis.

Mitosis is usually a controlled process. For example, the rate of mitosis in the epidermis of the skin is increased when the skin has been cut. Once the damage caused by the cut has been repaired, however, the rate of mitosis is decreased again. Sometimes, abnormal cells divide by mitosis in an uncontrolled way. This gives rise to tumours and, if the cells are malignant, cancers.

Mitosis consists of two processes: division of the nucleus (**karyokinesis**) and division of the cytoplasm (**cytokinesis**). Division of the cytoplasm usually occurs straight after division of the nucleus. Sometimes division of the cytoplasm is delayed. This leads to cells that have more than one nucleus. Cells in the muscles that move your skeleton are a rare example of human cells that have many nuclei.

Although mitosis is a continuous process, it is usually described as a series of stages. These stages are prophase, metaphase, anaphase and telophase. The time between divisions is called interphase. Table 9.1 summarises the main events that take place during each stage of mitosis. This gives us an overview of mitosis. We will then look at each stage in more detail.

Q 2 Name three processes in humans that involve mitosis.

Stage of mitosis	Main events that occur
Prophase	Chromosomes start to coil, becoming shorter and fatter Nuclear envelope disappears A network of protein fibres (the spindle) forms in the cell
Metaphase	Chromosomes line up on the equator of the spindle Fibres of the spindle attach to a region of each chromosome called the centromere
Anaphase	Spindle fibres contract and pull the two copies of each chromosome to opposite ends of the spindle
Telophase	The two sets of chromosomes form new nuclei During this process, the chromosomes become long and thin again and new nuclear envelopes form around the nuclei The cytoplasm usually divides to form two new cells
Interphase	The daughter cells grow more cytoplasm and get on with their normal activities The cells make copies of their chromosomes (by DNA replication) ready for their next division

Table 10.1 A summary of events that occur during the stages of mitosis.

Figure 10.1
A cell during interphase, i.e. prior to mitosis occurring.

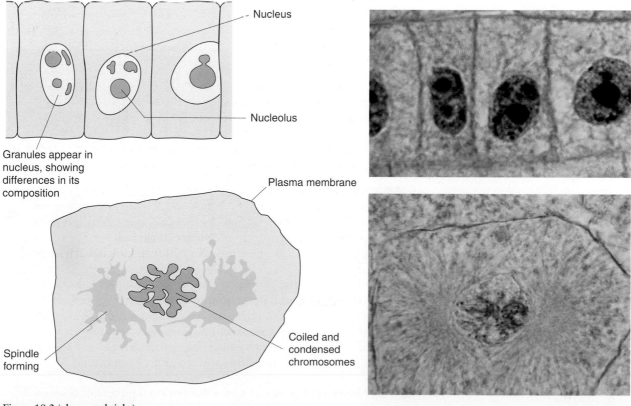

Nucleus

Nucleolus

Granules appear in nucleus, showing differences in its composition

Plasma membrane

Spindle forming

Coiled and condensed chromosomes

Figure 10.2 (above and right)
Cell during prophase of mitosis.

key term

Chromatids are the two copies of a single chromosome held together by a centromere. They are made by DNA replication and are exact copies of the original chromosome in the parent cell.

Figure 10.3
Electronmicrograph of replicated chromosomes.

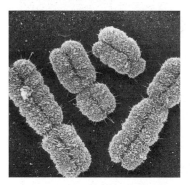

Prophase

Chromosomes are normally long and thin. They are so thin that they cannot be resolved by an optical microscope. This is why we cannot see them in the cell shown in Figure 10.1.

During prophase, chromosomes get shorter and fatter. They do this by coiling on themselves, a process called **condensation**. As the chromosomes condense, they become visible using an optical microscope. Figure 10.2 shows a cell during prophase. You can see its chromosomes. Each chromosome has two strands. These strands are the two copies of the original chromosome. They are held together somewhere along their length by a structure called a **centromere**. As long as the two copies of a chromosome are held together by a centromere, they are each called a **chromatid**. The chromatids of a single pair are called **sister chromatids**.

Q 3 Cells from whitefish were chosen for Figure 10.2 because this organism has relatively few chromosomes. This makes them easier to see. How many chromosomes are present in a human cell?

Figure 10.3 is an electronmicrograph of replicated chromosomes. In this photograph you can see the sister chromatids and centromeres very clearly.

You can see the spindle in Figure 10.2. This spindle is made from fibres of protein that radiate from two poles. During prophase, fibres from the spindle become attached to each side of the centromeres holding pairs of sister chromatids together.

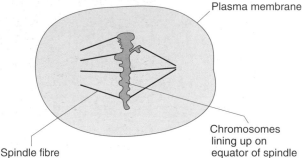

Plasma membrane

Chromosomes lining up on equator of spindle

Spindle fibre

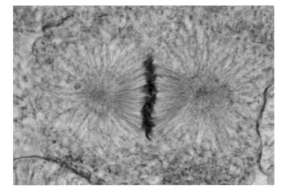

Figure 10.4
A cell during metaphase of mitosis.

Metaphase

In Figure 10.2, the chromosomes are dispersed throughout the cell. Compare this with the cell in metaphase shown in Figure 10.4. The chromosomes in Figure 10.4 have moved to the middle of the spindle. This distinguishes a cell in metaphase: its chromosomes are lined up at the equator of the spindle. We cannot see that spindle fibres have attached to the centromeres of the sister chromatids of each chromosome.

Q 4 What has happened to the nucleus of the cell shown in Figure 10.2?

Q 5 Figure 10.4 shows us a two-dimensional picture of the spindle. Suggest what its three-dimensional shape will be.

Anaphase

Two important events occur during anaphase:

- the centromeres divide into two: this separates the sister chromatids in each chromosome

- the fibres of the spindle contract: this pulls the two sister chromatids apart in opposite directions, eventually reaching opposite poles of the spindle.

Figure 10.5
A cell during anaphase of mitosis.

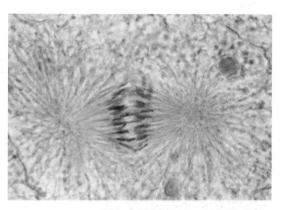

Pole of spindle

Spindle fibre

Chromatid being pulled to left of cell

You cannot see the movement of chromatids in Figure 10.5. However, you can see its effects: the chromatids are now separate and are no longer at the equator of the spindle.

Plasma membrane beginning to 'pinch' cell in two

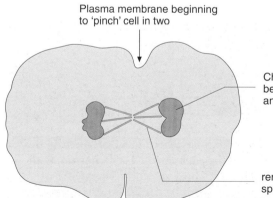

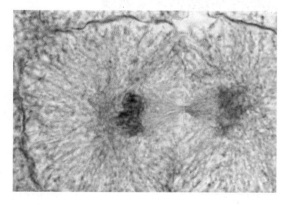

Chromosomes becoming long and thin again

remaining spindle fibres

Figure 10.6
A cell during telophase of mitosis.

Telophase

One chromatid from each chromosome reaches the poles of the spindle. Around each pole there is now an exact copy of each chromosome that was present in the nucleus of the parent cell. Nuclear division is complete. This is telophase, shown in Figure 10.6. Once they have reached the poles of the spindle, the chromosomes (as we can now call them again) revert to their original form. This means that they become long and thin.

Q 6 Suggest one advantage to explain why chromosomes:

    (a) are normally long and thin

    (b) become short and fat prior to mitosis.

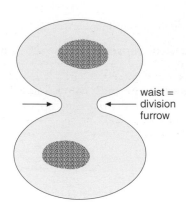

Figure 10.7
Division of the cytoplasm
(cytokinesis) in an animal cell.

Cytokinesis

Except in cells such as the muscle cells that attach to your skeleton, division of the cytoplasm follows telophase fairly quickly. Figure 10.7 shows how a 'waist' forms in the middle of the cell. Eventually, plasma membrane from one side of the cell joins that of the opposite side of the cell and the two new cells separate.

Studying mitosis in plant tissues

You cannot examine mitosis in human cells in your school or college laboratory. Instead, you will examine mitosis in plant cells. Figure 10.8 shows cells from the root tip of an onion. Plant tissue was chosen because it can be used safely in a school or college laboratory. A root tip was chosen because its cells divide rapidly. You can see that many of the cells in Figure 10.8 were dividing by mitosis. Table 10.2 describes the steps you would take to produce a temporary mount of onion cells, like that shown in Figure 10.8.

Figure 10.8
Cells from the root tip of an onion. Many of these cells are dividing by mitosis.

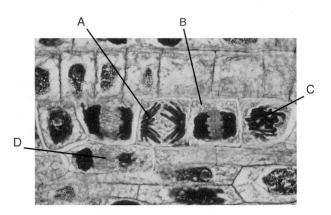

Q 7 Some of the cells in Figure 10.8 have been labelled. Identify the stage of mitosis represented by each labelled cell. Starting with prophase, arrange the labels to show the correct sequence of stages in mitosis.

Q 8 Which phase, or phases, of the cell cycle corresponds to interphase?

Steps in procedure	Explanation
Use a scalpel to cut about 3 mm from the tip of a growing root	Dividing cells are found only at the very end of the growing root
Put the cut root tip onto a watch glass	Behind the root tip, cells have elongated and differentiated into mature, non-dividing cells
Wearing gloves, add 1 mol dm^{-3} acetic acid (also known as ethanoic acid) to the root tip	The warm acid helps to soften the tissues so that cells can be separated more easily
Put the watch glass onto a hotplate and warm for about 5 minutes (do not let the acid boil)	
Use forceps to put the root tip onto a glass slide	We always mount specimens on a glass slide before viewing them with an optical microscope
Add a few drops of aceto-orcein stain to the root tip on the slide	The stain is taken up by chromosomes, making them easier to see
Gently break up the root tip using a mounted needle	This separates cells
You should attempt to spread the root tip out thinly, rather than stir the cells from the root	
Put a coverslip over the root and cover the coverslip with a piece of filter paper	The aim here is to produce a layer that is only one cell thick
Press down on the coverslip gently with your thumb (if you press too hard, the coverslip will break)	This makes individual cells easier to see
Use an optical microscope to examine the slide for cells dividing by mitosis	You will now see onion cells that have their chromosomes stained a reddish purple colour

Table 10.2 Steps in the production of a root tip squash. This process enables us to see cells dividing by mitosis.

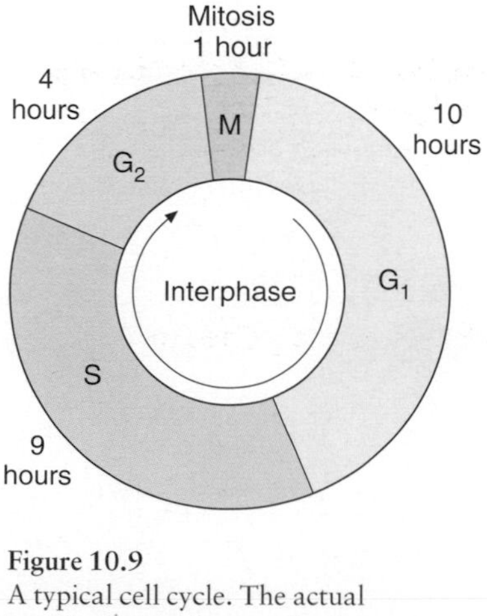

Figure 10.9
A typical cell cycle. The actual length of each phase varies from one organism to another.

The cell cycle

Like all eukaryotic cells, human cells have a well-defined cycle. In rapidly dividing cells, the cycle usually lasts for several hours and ends with mitosis. Figure 10.9 represents the cell cycle. Each phase of the cycle involves specific cell activities:

- **G1**: cells prepare for DNA replication
- **S**: DNA replication occurs
- **G2**: a relatively short gap before mitosis
- **M**: mitosis.

You can see from Figure 10.9 that mitosis is a relatively short part of the cycle.

Summary

- Cells can be made only when existing cells divide.

- There are two types of cell division in eukaryotic cells (such as human cells): mitosis and meiosis. Cell division in prokaryotic cells (such as bacteria) does not involve mitosis or meiosis.

- Mitosis occurs in humans during growth, during the replacement of lost cells and during repair of damaged tissues. Some mature cells, e.g. nerve cells, lose their ability to divide by mitosis. Tumours and cancers result from the uncontrolled mitosis of abnormal cells.

- Mitosis produces cells that are genetically identical to each other and to the parent cell.

- Mitosis consists of several stages: prophase, metaphase, anaphase and telophase. Interphase is the period between mitotic divisions.

- In prophase, chromosomes condense to become shorter and fatter. We can see them using an optical microscope. Each chromosome has already been copied. The two copies, called sister chromatids, are held together by a centromere. The nucleus disappears and a spindle of protein fibres forms during prophase.

- In metaphase, the chromosomes line up on the equator of the spindle.

- In anaphase, each centromere divides, separating sister chromatids. Contraction of fibres in the spindle pulls the two sister chromatids of each pair to opposite poles of the spindle.

- In telophase, the separated chromatids (now called chromosomes again) uncoil to become long and thin again. Division of the cytoplasm (cytokinesis) usually occurs. In human cells this is done by pinching the cytoplasm in the middle of the cell.

- Interphase is the time between cell divisions.

- Dividing onion cells can be viewed by preparing a temporary mount of squashed root tip, stained using aceto-orcein.

- Rapidly dividing cells undergo a cell cycle. The G1, S and G2 phases of this cycle represent interphase. The M phase of the cycle represents mitosis.

- DNA replication occurs in the G1 phase of the cell cycle.

Assignment

There are many complex techniques that can be used to investigate the processes which occur in living organisms. Biologists can also find out a lot about cell biology by using simple instruments such as light microscopes and interpreting what they see very carefully. This assignment will provide you with an opportunity to interpret a photograph showing cells undergoing mitosis. The photograph was taken through an ordinary light microscope.

Figure 10.10
Cells at various stages in mitosis.

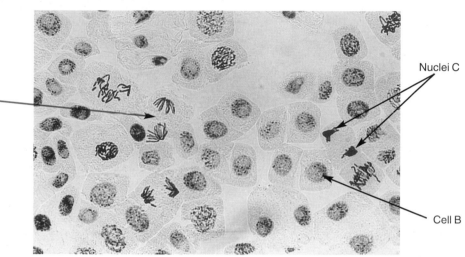

Look at the photograph in Figure 10.10. Some of the cells which you can see are dividing.

1 What is the evidence from the photograph that:

   (a) these are animal cells?

                    (1 mark)

   (b) the nucleus labelled A is from a cell in anophase?

                    (1 mark)

2 (a) Measure the horizontal width of cell B in millimetres.

   (b) The magnification of this photograph is ×250. Use this and the measurement from Question 2(a) to calculate the real diameter of the cell. Give your answer in micrometres (μm). Show your working.

                    (2 marks)

3 Cell B is a cell which is in interphase. It is just about to divide. The nuclei labelled C are in telophase. The nucleus of cell B contains 14 chromosomes and 10 units of DNA. How many chromosomes and how much DNA would you expect to find in each of the nuclei labelled C?

                    (2 marks)

4 Like other animal cells, human cells also undergo mitosis. Suggest whether or not you would expect to see cells undergoing mitosis in slides of each of the following. Give a reason for your answer in each case.

(a) Red blood cells.
(b) Cells from the lining of the intestine.
(c) Cells from the surface of the skin.

(3 marks)

Table 10.3 shows the number of cells in interphase and in the various stages of mitosis in this photograph

Stage	Number of cells	Percentage of cells
Interphase		
Prophase		
Metaphase		
Anaphase		
Telophase		
Total		

Table 10.3 Number of cells in interphase and stages of mitosis.

5 (a) Use information from the photograph to complete column 2 of this table as accurately as you can. Do not worry if you cannot decide exactly what stage every cell is in.

(1 mark)

(b) Calculate the percentage of cells in each of the stages. Write your answers in the third column of the table.

(1 mark)

6 (a) Which stage of mitosis would you expect to take longest? Explain how you arrived at your answer.

(1 mark)

(b) It takes 24 hours for a cell from this animal to go from the beginning of prophase to the end of telophase. Assume that the percentage of cells in a particular stage of mitosis is proportional to the time taken for the cell to pass through that stage. Use the information in the table to estimate the lengths of the various stages of mitosis. Present your results as a suitable scale drawing.

(2 marks)

Examination questions

1 (a) Name the phase in the cell cycle in which the following events take place.

    (i) DNA replicates

    (ii) Copies of each chromosome move to the poles of the cell.

(2 marks)

(b) Figure 10.11 shows a section through an onion bulb that is starting to grow.

    (i) Which of the parts labelled A to D of this bulb would you use to prepare a slide showing mitosis?

    (ii) Name the stain you would use to make this preparation.

    (iii) Describe the effect this stain would have on the cells.

(3 marks)

Figure 10.11

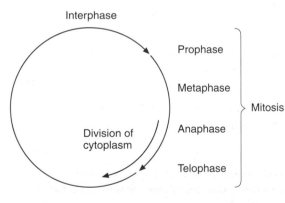

2 Figure 10.12 shows the main stages in the cell cycle.

(a) At what stage in the cell cycle do the following events take place?

    (i) Shortening of the chromosomes.

    (ii) Replication of DNA.

(2 marks)

(b) The root tip of a growing plant was used to prepare a slide to study mitosis.

    (i) Name one stain that you could use to make this preparation.

(1 mark)

    (ii) The number of cells at each stage of the cell cycle were counted. Table 10.4 shows the number of cells at each stage.

Stage of cell cycle	Number of cells seen on slide
Interphase	147
Prophase	68
Metaphase	10
Anaphase	4
Telophase	15

Table 10.4 Calculate the percentage of time in the cell cycle spent in metaphase. Show your working.

(2 marks)

Figure 10.12

3 During the twentieth century, several experiments were carried out to discover the chemical nature of genetic material.

(a) Alfred Mirsky found that all the body cells of any species contain the same amount of DNA, which is double the amount found in egg and sperm cells. However the same was also true of protein.

What conclusion can you make from this evidence about the chemical nature of the genetic material?

(1 mark)

(b) In 1928, Frederick Griffith was interested in developing a form of vaccine against a fatal form of pneumonia, caused by the bacterium *Pneumococcus*. This bacterium has two forms: one with an outer capsule and one without the outer capsule. He treated the two forms of these bacteria in different ways and then injected them into mice. Some of his results are shown in Table 10.5.

Type of *Pneumococcus* injected into mice	Effect of injection
Live, non-capsulated	Mice remained healthy
Live, capsulated	Mice died of pneumonia
Heat-killed, capsulated	Mice remained healthy
Heat-killed, capsulated and live non-capsulated	Mice died of pneumonia

Table 10.5

(i) Which type of *Pneumococcus* causes pneumonia? Explain your answer.

(1 mark)

(ii) Suggest why the mice died of pneumonia after being injected with heat-killed, capsulated *Pneumococcus* and live, non-capsulated *Pneumococcus*.

(2 marks)

(c) Bacteriophages are a type of virus that infect bacteria. The T_2 bacteriophage has an outer protein coat surrounding its DNA core. When the T_2 bacteriophage infects the bacterium *Escherichia coli*, it injects its own DNA into the bacterial cell but leaves its protein coat on the surface of the bacterium. In 1952, Hershey and Chase performed an experiment in which they labelled the outer protein coat of T_2 bacteriophage with radioactive sulphur and labelled its DNA with radioactive phosphorous. They then infected cells of *Escherichia coli* with these T_2 bacteriophages. Table 10.6 summarises their results.

Part of T_2 bacteriophage labelled	Site of radioactivity in *Escherichia coli* after infection with labelled bacteriophage
Outer coat	Outside bacterial cell
DNA core	Inside bacterial cell

Table 10.6

(i) What was the source of radioactivity inside the cells of *Escherichia coli*?

(1 mark)

(ii) Explain what these results suggest about the chemical nature of the genetic material.

(3 marks)

Gene Technology

Collagen is the most common protein in humans, making up almost one-third of all protein in our bodies. It is a fibrous protein that provides structural support for blood vessels, bone, skin and tendons. Collagen is used by surgeons to prevent excessive bleeding and for suturing and repair. It is also used in cosmetic surgery to improve the appearance of scars.

Until recently, collagen was obtained from two major sources: cadavers (dead humans) and cows. However, the emergence of acquired immune deficiency syndrome (AIDS) in humans and of bovine spongiform encephalopathy (BSE) in cattle has raised fears about the possible contamination of these sources. This has triggered research to find sources of collagen that are free from these hazards.

Two yeasts appear to offer a solution. A team of researchers in The Netherlands has found that the yeast *Hansenula polymorpha* can produce human collagen. Another team in Finland has found that the yeast *Pichia pastoris* can also produce human collagen. By introducing the human collagen gene into the DNA of these two yeasts, large-scale production of human collagen is possible. This collagen is unlikely to be contaminated with the infective agents that cause AIDS or BSE.

Overview of gene technology

In this chapter we will learn about the way in which fast-growing cells can be used to make a human protein. Because the process involves combining human DNA with the DNA from another organism, it is called recombinant DNA technology. Figure 11.1 gives an overview of the process. We will examine each step in this flowchart in the rest of this chapter.

Figure 11.1
Using recombinant DNA technology to make a human protein.

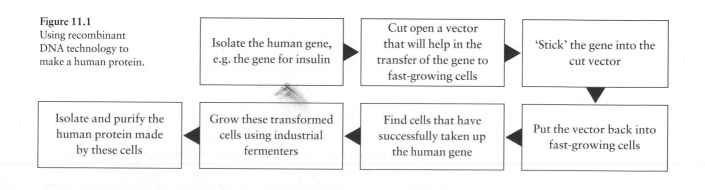

Isolate the human gene, e.g. the gene for insulin → Cut open a vector that will help in the transfer of the gene to fast-growing cells → 'Stick' the gene into the cut vector → Put the vector back into fast-growing cells → Find cells that have successfully taken up the human gene → Grow these transformed cells using industrial fermenters → Isolate and purify the human protein made by these cells

key term

Recombinant DNA technology involves the combination of DNA from one organism with DNA from another organism. Often this involves inserting human DNA into the DNA of another organism. When these genetically engineered organisms are cultured, they produce human protein.

key term

Reverse transcriptase is an enzyme that controls the formation of cDNA from mRNA.

Isolating the required gene

Except for egg cells and sperm cells, every cell in the human body contains 46 chromosomes. Each chromosome can contain several thousand genes. The first task is to isolate the gene that we want to use. There are three ways in which this can be done and these are described below. The first two ways depend on base pairing. If you have forgotten this process, it would be a good idea to refresh your memory by revising Chapter 9 before reading the rest of this chapter.

Working backwards from the protein

This is, perhaps, the easiest of the three methods. First we must work out the amino acid sequence of the protein we wish to make, e.g. insulin. Once we know this, we can use the genetic code in Table 9.3 to work out the base sequence that would code for this protein. We then make DNA with that base sequence. The result is an artificial gene made from complementary DNA (cDNA).

Using messenger RNA

When cells make a protein, they first transcribe its gene into a molecule of mRNA (see Chapter 9). Messenger RNA molecules carrying the code for insulin are common in the cytoplasm of insulin-producing cells. If we can find molecules of insulin mRNA, we can use them to make artificial insulin genes. To do this, we must use an enzyme called **reverse transcriptase**. This enzyme speeds up the production of cDNA from mRNA:

$$\text{mRNA + DNA nucleotides} \xrightarrow{\text{reverse transcriptase}} \text{cDNA + mRNA}$$

Using DNA probes to find our gene

To find a single gene amongst the many millions of genes in a human cell is a difficult task. To help this process, we first make a **DNA probe**. A DNA probe is a short, single strand of DNA that carries part of the base sequence of the gene we are looking for. The DNA probe is labelled with a radioactive or fluorescent marker. In the right conditions, molecules of the DNA probe will attach to the complementary sequence of bases in DNA extracted from human cells. The radioactive or fluorescent marker shows us where the target gene is.

Q 1 What would be the base sequence of a DNA probe used to find the sequence ATC GAC CCT AGA?

Cutting the gene out of its DNA chain

Once we have found the target gene in this way, we must remove it from its chromosome. Enzymes called **restriction endonucleases** (restriction enzymes for short) control this process. There are many different restriction enzymes. Each cuts DNA at a different base sequence, called a recognition sequence. Figure 11.2 shows the action of two restriction enzymes.

key term

Restriction endonucleases are enzymes that 'cut' DNA at specific base sequences within their molecules. They are made naturally by bacteria, which use them to destroy the DNA from the bacteriophages that infect them. These enzymes are usually called **restriction enzymes** for short.

Figure 11.2
The action of two different restriction enzymes. Note that the recognition sequence of each enzyme is six pairs long. Recognition sequences for restriction enzymes are usually four to eight base pairs long. Note also that each recognition sequence is palindromic, i.e. it reads the same in both directions.

Notice that the recognition sequence for each enzyme is palindromic, i.e. it reads the same in both directions. Figure 11.2 shows that one restriction enzyme, *Sma* I, cuts DNA vertically. This produces two fragments of DNA with blunt ends. The other restriction enzyme, *Pst* I, cuts DNA horizontally as well as vertically. This produces protruding ends. These protruding ends will form base pairs with any piece of DNA with the complementary sequence. Because of this, they are known as cohesive or '**sticky**' **ends**.

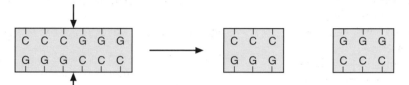

The restriction enzyme *Sma* I cuts DNA at the points shown by the arrows to form blunt ends

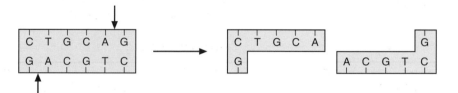

The restriction enzyme *Pst* I cuts DNA at the points shown by the arrows to form 'sticky ends'

Q **2** Identify the role of:

(a) reverse transcriptase
(b) restriction endonuclease.

Splicing a gene into a vector

The section above describes how we can isolate the gene that codes for a human protein we want to make. We next need to insert it into the type of cell we will use to make the protein. The easiest way to do this is to use a **vector**. Often we use a **plasmid** as a vector. Plasmids are short, circular strands of DNA that are found in some bacteria. We could also use bacteriophages (viruses that infect bacteria) as vectors.

key term

A **vector** is gene carrier. It will carry a human gene into the cell of a bacterium or yeast that will be used to make human protein. **Plasmids**, circular strands of DNA found in some bacteria, are useful as vectors if bacteria are to be used to make human protein.

To use a plasmid as a vector, we must cut it using the same restriction enzyme that we used to cut the human gene. Using the same restriction enzyme is very important. We want to produce complementary sticky ends on the human gene and the plasmid. Only then will they pair together.

Q 3 Why must the 'sticky ends' of a target gene and a plasmid be cut using the same restriction endonuclease?

Cut genes and cut plasmids are mixed together. Under the right conditions, the sticky ends of the gene and the sticky ends of the plasmid join together. This repair process is known as ligation. It is controlled by an enzyme called **ligase**. Figure 11.3 shows that the new molecule produced is a circular DNA molecule. Because it contains the genes that were in the original plasmid and the human gene, it is called **recombinant DNA**.

Figure 11.3
'Sticky ends' of cut DNA will join together.

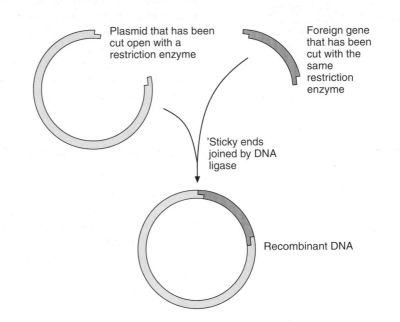

Introduction of recombinant DNA to the host cell

The recombinant DNA must now be introduced to a host cell. Several successful methods have been found by trial and error. One involves soaking bacteria in an ice-cold calcium chloride solution containing the recombinant DNA plasmids. After incubation at 42°C for two minutes, many bacterial cells take up the recombinant plasmids. Because these bacteria contain DNA from another species (humans) they are **transgenic organisms**.

Finding the genetically modified bacteria

Whatever method is used, not all the bacteria will take up the recombinant plasmid. Since the bacteria are useful to us only if they have the recombinant plasmid, we need to find them. Genetic markers are used in this process.

Q 4 The β-lactamase gene (called *amp*ʳ) confers resistance to the antibiotic ampicillin. Suggest why the *amp*ʳ gene might be inserted into a plasmid that we are hoping will carry the gene for insulin production into a bacterial cell.

The most common genetic marker is a gene for antibiotic resistance. Using antibiotic resistance as a genetic marker involves splicing two genes into plasmids: one for the human protein we want to make and a second for antibiotic resistance. If a bacterial cell has successfully taken up the plasmid with the human gene, it will also be resistant to the antibiotic. Cells that have not taken up the plasmid will be susceptible to the antibiotic. When bacteria are grown on plates of agar, they form distinct colonies. If the agar plate contains an antibiotic, only those bacteria with a gene for resistance will form colonies.

Culturing host cells

The above processes sound easy. They are not. Obtaining bacterial cells that have successfully taken up the modified plasmid is a hit-and-miss process. In the majority of cases, it will turn out to be unsuccessful (only about 0.0025% of treated bacteria successfully take up the plasmid). Even when bacteria pass the screening process, they do not always use the human gene they now contain. However, those that do are used to make pure cultures. These cultures are grown in large tanks called fermenters. Conditions inside the fermenters are carefully controlled to ensure the optimal growth of the transgenic bacteria. As they grow, the bacteria inside the fermenter make the target human protein. At appropriate intervals, the culture solution is drawn off and the human protein isolated from it.

Uses of recombinant DNA technology

Humans have used organisms for thousands of years. Plants were used for shelter, as tools, as food and as a source of medicine. Yeasts were used to make bread and alcohol. Bacteria were used to make cheese, yoghurt and vinegar. Animals were used for food and clothing as well as for working with humans. As farmers, humans took care of these organisms to ensure that they gained a large crop or kept animals that were strong. However, humans also genetically changed these organisms. Only those organisms that showed the best characteristics were allowed to breed. As a result, crop plants were bred that gave a large crop. Animals were bred to give a high yield of meat, milk or wool. This process is called **artificial selection**. It is called 'artificial' because it is humans, and not nature, that determine which characteristics dominate the population. The animals shown in Figure 11.5 show some results of artificial selection. The sheep have been bred over the centuries to produce more wool than their natural ancestors. The female cattle have been bred to have a high yield of milk. The sheep dog is only one of a variety of breeds of dog.

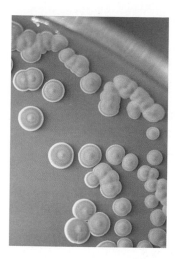

Figure 11.4
When cultured on agar plates, individual bacteria divide to form colonies. Each colony you can see on this agar plate contains millions of cells. The cells in each colony were produced in only a few hours by the division of a single cell.

Figure 11.5
Breeds of animals produced by artificial selection. (a) A merino sheep with short legs, to stop it escaping from pens, and long hair, for a high yield of wool. (b) Fresian cattle have a very high milk yield. (c) The sheep dog has been bred to round up sheep without harming them.

(a)　(b)　(c)

Recombinant DNA technology allows us much greater control over genetic manipulation. Table 11.1 (overleaf) summarises some of the ways that recombinant DNA technology has been beneficial to humans.

Application of recombinant DNA technology	Explanation
Genetically engineered microorganisms	Human genes can be inserted into bacteria, which are then grown in fermenters The bacteria produce a human protein Large amounts of insulin and human growth hormone can be produced cheaply in this way
Genetically modified plants	Desirable genes can be transferred from an organism to a crop plant Potato plants that are resistant to attack by a virus and maize (corn) plants that are resistant to drought have been produced in this way
Genetically modified animals	Human proteins, such as haemoglobin and blood-clotting factors, are already produced in the milk of transgenic cows, goats and sheep
Human gene therapy	Gene therapy involves inserting a 'normal' gene into an organism's body to correct a genetic disorder Severe combined immune deficiency (SCID) leaves some people with almost no functioning immune system Inserting copies of a gene coding for the enzyme adenosine deaminase (ADA) into the blood cells of sufferers has been used to 'cure' the symptoms of SCID
Mapping human chromosomes (the Human Genome Project)	Scientists in over 1000 laboratories around the world are contributing to the Human Genome Project Their aim is to create a map of all human chromosomes, identifying the precise location of every gene This will help to develop new gene therapy treatments

Table 11.1 Some applications of recombinant DNA technology that have benefitted humans.

Q 5 Explain what is meant by *transgenic cows* in Table 11.1.

Moral and ethical concerns about recombinant DNA technology

Over the past few years, there have been a number of demonstrations about the growth of genetically modified organisms (GMOs). National newspapers have covered these demonstrations, and have also carried articles debating the use of GMOs. Some stories seem designed to upset readers. For example, one newspaper in the UK ran an article suggesting that if you ate meat from an animal with an inserted human gene, you would be a cannibal. You should be able to distinguish these stories from rational concerns about GMOs.

Over 70% of the land area of the UK is under some form of agricultural production. None of this is currently used for the commercial growth of GMOs. Research into the possible use of genetically modified crop plants is, however, underway in many countries. Table 11.2 shows some of the ways in which genetic engineering could, in principle, be used to produce novel crops. However, there is international agreement that genetically modified crops should pass through a safety assessment procedure before they can be grown. This involves careful research to answer the following questions.

- How does the introduced gene affect the genetically engineered plant?

- Is there evidence that the introduced gene affects the toxicity or allergenic properties of the genetically engineered plant?

- Is it likely that there will be unintended effects on other, useful organisms within the environment?

- Is the genetically engineered plant likely to become a weed or to invade natural habitats?

- Can the introduced gene be transferred to other plants (e.g. by pollination) or animals and, if so, what would be the likely consequences?

Examples of ways in which genetic modification could make crops more useful
Improve resistance to pests
Improve resistance to bacterial, fungal or viral diseases
Improve resistance to herbicides, so that crop plants are unaffected by herbicides used to control weeds
Remove allergens from crops that currently contain them, e.g. rice
Enable plants to resist cold or drought
Produce salt-tolerant varieties of crops – important in parts of the Indian subcontinent and sub-Saharan Africa if global warming causes a rise in sea level
Improve storage time, e.g. by slowing the sprouting of potatoes
Enable plants to produce pharmaceutical substances, such as edible vaccines
Enhance the production of vitamins and anti-cancer substances by plants

Table 11.2 In principle, it is possible to produce genetically engineered crops with novel properties.

Despite the safety checks, many people, including scientists and politicians, are concerned about the risks of recombinant DNA technology. Table 11.3 (overleaf) summarises some of their concerns.

Moral or ethical concern	Explanation
Mutation of transgenic bacteria or viruses	If they mutate, these transgenic organisms could become new pathogens The danger is that we might not be able to control these new pathogens
Genetically modified crops could 'escape'	If pollen or seeds were carried from test plots, they might result in genetically modified populations elsewhere Since the genetic modification could involve resistance to herbicides, these plants could become 'superweeds' that we could not control
Transgenic organisms might set up an evolutionary process that could harm the environment	Some crop plants have been given a gene enabling them to produce a pesticide Through natural selection, this might speed up the evolution of pesticide-resistant insects
Populations of transgenic organisms could upset the balance of nature	Populations of transgenic salmon have been produced in which individuals grow rapidly These transgenic fish could compete for food with other fish species The latter could become extinct and natural food webs could change
Objection to specific transgenic organisms	Many religious groups could not use products from specific organisms, e.g. to Hindus, cows are sacred animals and to Jews and Muslims, pigs are unclean The use of products from these organisms might be unacceptable to people from these religions
Eugenics	Genetic engineering could be used to insert genes into humans (or their eggs and sperm) In some cases, for example SCID, this process could prevent an early death However, it could be used to give individuals characteristics that are considered to be desirable This is unacceptable to many people and it also reminds people of some of the programmes that have been used throughout modern history to eradicate less powerful ethnic groups
Screening could lead to discrimination against individuals	When a fetus is screened for genetic disorders, parents often face a dilemma about aborting an affected fetus It will become possible to screen adults for genes that predispose them to genetic disorders. This might lead to insurance companies discriminating against people with these disorders, even though they have perfect health

Table 11.3 Many people are concerned about the use of recombinant DNA technology. This table summarises their concerns.

Q 6 Suggest how eugenics could be used to help sufferers of inherited disorders such as cystic fibrosis (described in Chapter 9).

Extension box

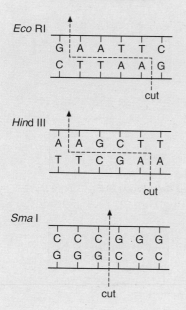

Figure 11.6
The base sequences recognised by three different restriction enzymes: *Eco* RI, *Hin*d III and *Sma* I. The sequences recognised by many restriction enzymes are six base pairs long and the sequence is the same whichever way round it is read.

It may surprise you to know that humans are not the only organisms to suffer from viral infections. Bacteria can also be infected by viruses.

Viruses have a very simple structure. They consist essentially of a protein coat surrounding a piece of nucleic acid which codes for the proteins in this coat. When a virus in which the nucleic acid is DNA infects a bacterial cell, its DNA enters the bacterium and takes over the host cell's metabolism. As a result, the bacterium makes new virus particles instead of its own proteins.

Many bacteria produce enzymes which recognise foreign DNA like that from viruses and disable it by cutting it into short pieces. These enzymes are called restriction enzymes because their activity is restricted to foreign DNA. Each enzyme recognises a specific sequence, often six nucleotides long, and only cuts the DNA at this site. Figure 11.6 shows some different restriction endonucleases and the nucleotide sequences where they cut the DNA. More than a hundred different sorts of restriction enzyme have now been isolated and purified. They get their names from the bacteria from which they are obtained. *Eco* RI, for example, was the first restriction enzyme (R I) isolated from *Escherichia coli* (*Eco*).

If you look at Figure 11.6, you will see that *Sma* I cuts the length of DNA in a different way to the other two. It cuts straight through the DNA producing two blunt ends. The other two enzymes produce staggered cuts sometimes called 'sticky ends'. Enzymes which produce staggered cuts are particularly useful in genetic engineering, enabling an introduced piece of DNA to join to the DNA in a plasmid.

Different restriction enzymes which cut the DNA at different points provide genetic engineers with a very useful set of tools. We can illustrate their value by looking at an example. Figure 11.7 shows a length of DNA. Suppose we want to isolate the gene that is represented by the base sequence, shown in red, from this length. If we use the enzyme *Eco* RI, it will cut this sequence in two places, effectively cutting out the gene we want. If we use *Hin*d III, on the other hand, the bases that it recognises are within the part of the sequence that makes up the gene. It would not be suitable to use this enzyme.

Figure 11.7
A choice of tools. Careful choice of restriction enzymes allows a genetic engineer to isolate a particular gene.

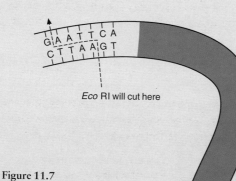

Eco RI will cut here

Eco RI will cut here

*Hin*d III will cut here. This would cut the gene into small pieces. We could not use this enzyme to get a copy of this gene

Summary

- Gene technology, also called recombinant DNA technology, involves combining DNA from one species with DNA from another species. It can be used to enable non-human organisms to produce human proteins, such as insulin.

- A desired gene can be made from its mRNA, obtained from cells, in other words by reversing the process of transcription. A reverse transcriptase enzyme is used in this process.

- Alternatively, a desired gene can be 'cut' from DNA using restriction endonuclease enzymes. When these enzymes produce staggered cuts, 'sticky ends' are formed on the DNA. The advantage of these sticky ends is that they will bind by base pairing to identical sticky ends in a vector.

- DNA probes can be used to help us find a desired gene. DNA probes have a base sequence that is complementary to parts of the desired gene. They also have radioactive, or fluorescent, markers to help us locate them once they have bound to the desired gene.

- Once isolated, a desired gene is spliced into a vector. A vector is a gene carrier that helps us to get the desired gene into the target organism.

- If the target organism is a bacterial cell, plasmids are often used as vectors. Plasmids are small circular DNA molecules found in some bacterial cells. The desired gene and the plasmid are cut using the same restriction endonuclease enzyme, and then stuck together. A ligase enzyme helps the gene and the plasmid to stick together, or anneal.

- Usually, a gene for antibiotic resistance is also spliced into bacterial plasmids. Growing bacteria on a medium containing antibiotic enables us to identify which bacteria have successfully taken up the transgenic plasmid.

- Successfully transformed bacteria are grown in fermenters. The human protein they make, e.g. insulin, is drawn off and purified from the culture medium.

- Recombinant DNA technology can be used for the treatment of some human disorders. It can also be used to harvest human proteins from the milk of cows, goats or sheep and to improve the quality of crops.

- Many people, including scientists, are concerned about the use of recombinant DNA technology.

Assignment

Figure 11.8
Mussels are shellfish. They produce a protein glue which helps them to stick firmly to rocks on the seashore. This glue even sticks under water!

Different organisms produce different proteins. There are many different species of living organism and there are probably even more different proteins. The genes which encode some of these proteins have been isolated and used to produce genetically modified organisms. These organisms make specific proteins in large and useful amounts.

This assignment is based on a passage about a protein which is produced by mussels. This protein helps to attach these animals to rocks on the seashore.

The questions which follow should help you to understand a little more about how particular proteins are adapted to different functions as well as about some of the problems encountered by scientists in introducing genes from one species to another. A good understanding of the material in the passage will require you to draw on topics from various parts of your AS course. Showing how different topics link together is an important skill and one which you will need to develop if you continue with your biology course.

Read the following passage:

Walk along the seashore at low tide and you will see a variety of different organisms. Many of these organisms are sedentary and stay in one place. One of the major problems that they face is that they must avoid being washed away by pounding waves. This is where proteins such as mussel adhesive protein come in.

Mussel adhesive protein has an unusual primary structure. It contains a sequence of ten amino acids which is repeated many times. Unlike proteins such as enzymes and the carriers in plasma membranes, mussel adhesive protein does not have a distinct tertiary structure. It is always changing shape. This allows the protein molecules to squeeze into the small cracks and crevices which cover rock faces. Like man-made glues, mussel adhesive protein starts off as a liquid and sets into a solid. It is the repeated amino acid sequences which allow the protein glue to set firm. Each of the repeated sequences contains the amino acid, tyrosine. Chemical bonds form between the tyrosine in one sequence and the tyrosine in another. Not only do these bonds form between different parts of the same molecule, but they also form between different molecules. The chemical bonds bind the protein molecules firmly to each other and to the surface to which the mussel is attached.

Tests have shown that mussel adhesive protein does not result in the production of antibodies when introduced into the human body. Because of this, it could have many uses in medicine. But, before we can begin to consider ways of using it, we need a method of producing suitably large amounts. Attempts have been made to use genetically modified bacteria to produce mussel adhesive protein but researchers have encountered a problem. Bacteria are able to synthesise the protein, but only in an inactive form. It is thought that the reason for this is that, in a mussel cell, chemical changes take place in the endoplasmic reticulum which convert the inactive molecule to an active one. Scientists believe, however, that we can overcome this problem by using transgenic tobacco plants to produce mussel adhesive protein. Tobacco is a good choice. Not only will the necessary chemical changes take place but tobacco is a non-food plant and it has no close wild relatives in Europe.

Use your own knowledge and information in the passage to answer the following questions:

1 (a) Describe how the structure of an enzyme molecule differs from the structure of mussel adhesive protein. *(2 marks)*

 (b) How is the structure of each of these molecules related to its function? *(3 marks)*

2 (a) If mussel adhesive protein is to be used in medicine, it is important that it does not result in the production of antibodies when introduced into the human body. Explain why. *(1 mark)*

(b) Suggest two specific uses for mussel adhesive protein in medicine. *(2 marks)*

3 Figure 11.9 shows the sequence of amino acids which is repeated many times in each mussel adhesive protein molecule.

Figure 11.9
This sequence of amino acids is repeated many times in each mussel adhesive protein molecule. The different colours represent different amino acids. Only tyrosine has been named.

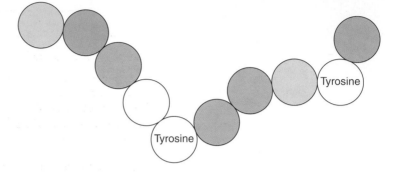

(a) How many mRNA nucleotides are necessary to encode this sequence? *(1 mark)*

(b) What is the minimum number of different sorts of tRNA necessary to bring the amino acids to the ribosomes to produce this sequence of amino acids? *(1 mark)*

(c) (i) Explain why we can work out a sequence of mRNA bases which encodes mussel adhesive protein if we know its amino acid sequence. *(1 mark)*

(ii) The sequence of mRNA bases for the adhesive protein actually found in mussel cells may differ from the sequence we have worked out. Explain why. *(1 mark)*

4 (a) The process converting inactive mussel adhesive protein into its active form is **post-translational modification**. Explain why. *(2 marks)*

(b) Explain how the organelles in a bacterial cell result in its being:
(i) able to synthesise mussel adhesive protein; *(1 mark)*
(ii) unable to produce the active form of the protein. *(1 mark)*

(c) Why would you expect transgenic tobacco plants to be able to produce active mussel adhesive protein? *(1 mark)*

5 Suggest why a plant chosen to produce mussel adhesive protein should:
(a) be a non-food plant; *(2 marks)*
(b) have no close wild relatives in the area where it is to be grown. *(2 marks)*

Examination questions

1 Figure 11.10 summarises stages involved in genetic engineering.

Figure 11.10

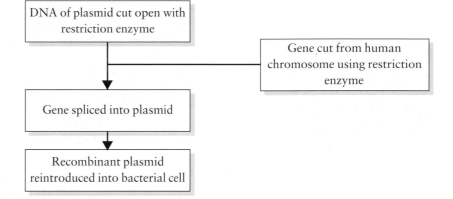

(a) Explain why the same restriction enzyme must be used to cut the plasmid and the human gene. *(2 marks)*

(b) Name the enzyme that is used to splice the human gene into the plasmid. *(1 mark)*

(c) Describe **one** method that can be used to discover bacterial cells that have successfully taken up the recombinant plasmid. *(3 marks)*

2 Clover is a crop plant that is fed to sheep. Genetic engineering has been used to develop strains of clover that have high concentrations of proteins rich in sulphur-containing amino acids. A piece of recombinant DNA was produced which contained three genes. The recombinant DNA, shown in Figure 11.11, was inserted into clover plants.

Figure 11.11

Gene 1 obtained from sunflower seeds. This gene codes for a protein rich in sulphur-containing amino acids

Gene 2 ensures that the protein rich in sulphur-containing amino acids is produced in leaf cells

Gene 3 prevents this protein being digested in the rumen of sheep

(a) Describe how enzymes could be used to remove Gene 1 from sunflower seeds and incorporate it into this recombinant DNA. *(3 marks)*

(b) It is hoped that feeding sheep on clover containing this piece of DNA will make them produce more wool. This is because wool is rich in sulphur-containing amino acids. Suggest why it is necessary to add genes **2** and **3** to make this possible. *(3 marks)*

Non-communicable Disease

The man in Figure 12.1 has suffered a heart attack. In fact, his heart has stopped beating. Hopefully, the electrodes being placed against his chest will stimulate his heart to begin beating again.

In 1997, at the age of 39, David Porter suffered a heart attack whilst scraping snow from his driveway at home. David says, "At first my arms felt weak but I thought that was just the cold. Later, when I was driving to work, I began to feel sick – really sick – so I stopped for a cup of tea. By this time, my arms felt like lead. I was struggling to breathe and sweating profusely. I felt as if a ten-ton truck had parked on my chest. Luckily, someone recognised my symptoms and called for an ambulance."

After an electrocardiogram and blood tests, it was confirmed that David had suffered a heart attack. He had two more aftershocks over the next four days he spent in the intensive care unit at his local hospital. Luckily, David survived his heart attack and returned home 11 days later. "When I think about it now," David says, "my health situation was awful. I stopped exercising when I gave up football. I ate lots of fatty foods and almost no fruit. I was working late into the night in a high-stress job as an operations manager for a telecommunications company. I weighed more than 15 stones and was always out of breath. You could say I was a heart attack waiting to happen."

David was clearly shaken by the experience. "You stop taking things for granted. Every twinge in your body frightens you. I have changed my lifestyle, though. I've lost over 3 stones and I'm back to the weight I was when I was 25. I eat no red meat, lots of vegetables and lots of fruit – at least five pieces a day. I walk three miles every morning, and three evenings a week I go for a swim at my local leisure centre."

David was lucky. In 1997, the year of his heart attack, over 76 000 men and 64 000 women died as a result of heart disease.

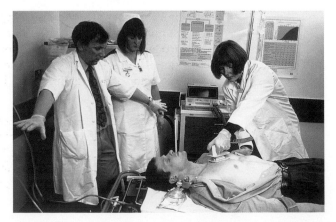

Figure 12.1
This man has suffered a heart attack. The electrodes provide an electric shock to re-start his heart.

In Chapter 7 you learned about diseases that are caused by pathogens. This chapter looks at diseases that result from our lifestyle. Figure 12.2 summarises the causes of death in the UK in 1997, the last year for which figures are available. You can see that the two major causes of death were coronary heart disease and cancer. We will look at these two 'lifestyle' diseases in this chapter.

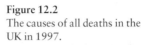
Figure 12.2
The causes of all deaths in the UK in 1997.

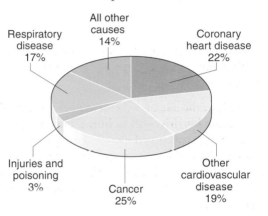

Coronary heart disease (CHD)

An adult human heart is about the size of a clenched fist and weighs about 350 g. It is made mainly of cardiac muscle (see Chapter 6). This muscle will contract over 2.5 billion times during the lifetime of an average, healthy human in the UK. During this time, it will pump blood through vessels whose total length is the equivalent of over 50 000 miles.

Atherosclerosis

To keep contracting like this, the cardiac muscle cells need a good supply of oxygen and glucose. This is supplied by blood. Although you might think that the heart is filled with blood, remember that the blood flows through the inner chambers of the heart. Like all other cells, cardiac muscle cells get their blood supply from branches of arteries: in this case, through the coronary arteries (Figure 12.3a). A blockage in a coronary artery will reduce the blood supply to cardiac muscle cells and interfere with their ability to contract.

- If a coronary artery becomes partly blocked, the cardiac muscle that it supplies will become short of essential nutrients and contract irregularly. Without enough oxygen to satisfy their needs, cardiac muscle cells get 'cramp'. This causes a pain in the muscle. Partial blockage of a coronary artery is often associated with a chest pain, called **angina**.

- If a coronary artery becomes totally blocked, the cardiac muscle that it supplies will die. Death to areas of cardiac muscle is called a **myocardial infarction**. Figure 12.3b shows how a blockage in a branch of a coronary artery will lead to myocardial infarction. If the area of dead cells is large, a myocardial infarction can be fatal.

Figure 12.3
(a) The heart muscle is supplied by coronary arteries. (b) Blockage of coronary arteries will lead to the death of heart muscle (myocardial infarction).

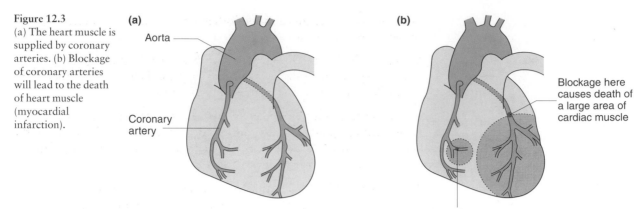

(a)

Aorta

Coronary artery

(b)

Blockage here causes death of a large area of cardiac muscle

Blockage here causes death of a small area of cardiac muscle

key term

Myocardial infarction
occurs when areas of cardiac muscle – the myocardium – die. This usually happens when the coronary artery supplying these cells becomes blocked. Myocardial infarction is commonly called a heart attack.

Q 1 (a) **What causes the pain felt by sufferers of angina?**

(b) **Suggest why angina might be an early warning of a heart attack.**

Use Figure 12.4 to remind yourself of the structure of an artery. It has three layers. The ones that interest us most are the inner, smooth endothelium and the middle layer of muscle cells. Notice that the artery in Figure 12.4 is healthy – it has a large, open lumen.

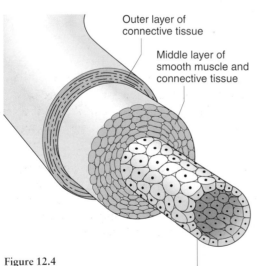

Outer layer of connective tissue

Middle layer of smooth muscle and connective tissue

Figure 12.4
A cross section through an artery.

Inner layer (endothelium)

key term

Atherosclerosis
is a disease in which the lumen of an artery becomes progressively narrower. It is often associated with an atheroma – a mound of fatty materials within the wall of the artery.

Atherosclerosis is a progressive disease that causes the narrowing of the lumen of an artery. No-one is really sure what causes atherosclerosis but there are two main theories.

● *Theory 1*   The endothelium of an artery becomes damaged. This might be caused by harmful chemicals, such as carbon monoxide, by viral infection or by a blow to the chest.

● *Theory 2*   In response to fatty materials leaking through the endothelium, phagocytic cells from the middle, muscular layer of the artery migrate to a new position just under the endothelium. Here they begin to divide rapidly, by mitosis, disrupting the endothelium.

Whatever the cause, once the endothelium has been disrupted the following events occur.

- Fats, in the form of lipoproteins, are taken up from the blood plasma and accumulate in the cells beneath the endothelium. This mound of swollen, fat-laden cells is called an **atheroma**.

- Muscle cells and fibres grow into the region of the atheroma, producing a **plaque**. As plaques grow, they bulge into the lumen of the artery (Figure 12.5a). This is atherosclerosis – the narrowing of the lumen of the artery.

- Eventually, the plaque can become so large that it tears the endothelium of the artery. As a result, a blood clot forms at the site of the tear (Figure 12.5b). A blood clot is also called a thrombus. Thus, the formation of blood clots within blood vessels is called **thrombosis**.

Figure 12.5
(a) A plaque forming inside an artery. (b) If the plaque damages the endothelium of an artery, a blood clot (thrombus) begins to form.

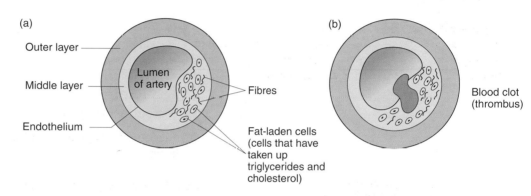

- A thrombus that breaks loose from its attachment is called an **embolus**. An embolus will travel in the blood until it is trapped in a small artery, which it completely blocks (an **embolism**). If an embolism occurs in a blood vessel to the brain, heart muscle or lungs, it can be fatal.

- Sometimes the build up in pressure that results from the narrowing of an artery causes the artery to swell and weaken. This is called an **aneurysm**. Swollen blood vessels press on surrounding tissues, often with harmful effects. Aneurysms often burst, leading to severe loss of blood (haemorrhaging). A brain aneurysm is called a **stroke**.

Q 2 Lipoproteins are formed when lipid (fat) molecules are combined with protein molecules. Use your knowledge of lipid structure to suggest why they are carried in the blood as lipoproteins.

Risk factors associated with coronary heart disease

Table 12.1 summarises some of the known factors that increase the risk of CHD. Note that many of them are interdependent. For example, lack of physical activity is likely to be associated with obesity, which itself carries an increased risk of diabetes mellitus, high blood pressure and high blood cholesterol levels.

Risk Factor	Notes
FACTORS YOU CANNOT AVOID	
• Increasing age	About 80% of people who die of CHD are aged 65 or older.
• Gender	Men have a greater risk of CHD than women and they have heart attacks earlier in life.
• Inheritance	People with parents who suffered CHD are more likely to develop it themselves. People of certain races are more likely to develop CHD, especially if they are in the minority in the country in which they live. This risk is probably stress-related and linked to high blood pressure (see below).
• Diabetes mellitus	Diabetes greatly increases the risk of CHD and of strokes. About two-thirds of diabetics die of some form of heart or blood vessel disease, making it critical that they work with their health care provider to control as many other risk factors as they can.
FACTORS YOU CAN AVOID	
• Tobacco smoke	Smokers are more than twice as likely to have a heart attack than non-smokers. Smokers who have a heart attack are more likely to die as a result than are non-smokers. Even exposure to other people's tobacco smoke – called environmental tobacco smoke or passive smoking – increases the risk of a heart attack.
• Blood cholesterol	The risk of CHD increases as blood cholesterol level rises. Female hormones tend to lower the total blood cholesterol level.
• Blood pressure	High blood pressure increases the heart's workload and causes it to weaken over time. High blood pressure also makes aneurysms – a ballooning out of the wall of an artery, more likely. A burst aneurysm leads to massive blood loss, a haemorrhage, which can be fatal. Stress and excessive consumption of alcohol increase blood pressure.
• Obesity	People who carry excess body fat are more likely to develop CHD even if they have no other risk factors.
• Inactivity	Over a long period of time, even moderate exercise can help to reduce blood cholesterol level, blood pressure and obesity.

Table 12.1 Risk factors associated with coronary heart disease (CHD).

We will examine three of the risks in more detail: tobacco smoking, high blood cholesterol and high blood pressure.

Tobacco smoking

Cigarette smokers are about four times more likely to develop CHD than non-smokers and are more likely to die of a heart attack when they have one. Pipe and cigar smokers are less at risk than cigarette smokers but are still about twice as likely to develop CHD than non-smokers.

The reasons for the risk associated with tobacco smoking are not clear. In some cases, they probably interact with other known risk factors listed in Table 12.1. The following mechanisms are thought to be involved.

● Carbon monoxide from tobacco smoke reduces the ability of red blood cells to carry oxygen which seems to increase the risk of damage to endothelial cells in arteries.

● Tobacco smoke contains nicotine, which is a vasoconstrictor. Vasoconstriction is another term for narrowing of arteries, which leads to increased blood pressure.

● Tobacco smoke is known to increase the cholesterol level in the blood. High cholesterol levels are a known risk factor.

Since CHD is only one of the effects of smoking tobacco, the healthiest strategy is not to smoke.

Blood cholesterol

Cholesterol is a type of lipid. When carried in the blood, lipids are combined with proteins, forming lipoproteins. Lipoproteins are divided into two categories:

● **low-density lipoproteins** (LDL). These contain high levels of cholesterol and increase the risk of developing atheromas

● **high-density lipoproteins** (HDL). These have low levels of cholesterol and do not contribute to atheromas. It is thought that HDL can actually mop up some of the cholesterol from LDL, making these LDL less dangerous to health.

Devices for testing blood cholesterol levels are cheaply available at pharmacists. Anyone with a blood cholesterol level above 240 mg per 100 cm^3 of blood is at high risk of developing CHD. People with a cholesterol level between 200 to 239 mg per 100 cm^3 of blood are considered to be borderline-high. It is therefore best to have a blood cholesterol level that is below 200 mg per 100 cm^3 of blood. Of this, cholesterol in LDL should be no higher than 130 mg per 100 cm^3 of blood; the rest should be in the form of HDL.

 3 Plasma membranes have receptors that are specific to certain groups of proteins. Suggest why LDL are more likely to cause the formation of atheromas than HDL.

High blood pressure

During vigorous exercise, blood pressure increases and contributes to an increased blood flow to active muscles. As you saw in Chapter 6, the increase in blood pressure is brought about by an increase in cardiac output and by contraction of the muscle in the middle layer of the arteries.

Short-term increases in blood pressure are normal. High blood pressure becomes a medical problem if it is permanently higher than normal in someone at rest. This person suffers from high blood pressure, or hypertension.

Over a long period of time, hypertension leads to an increased risk of developing CHD. It does this because an artery responds to increased blood pressure by increasing the thickness of its muscle layer. This causes its lumen to narrow, so increasing the blood pressure still further. Eventually, the blood pressure is so high that it is likely to damage the endothelium of the artery which, in turn, increases the risk of an atheroma forming.

Several factors lead to hypertension. These include stress, obesity, excessive consumption of alcohol, a high intake of sodium chloride and smoking tobacco.

Q 4 Suggest why people seeking a healthy diet are advised to:

(a) reduce their intake of animal fats;

(b) reduce the amount of salt they add to their food.

Cancer

Cancer is a major cause of death in the developed world. If it were a single disease with a single cause, cancer would probably be easier to cure. In fact, there are over 200 different cancers. Table 12.2 shows the occurrence of the more common cancers in the UK. The one thing these cancers have in common is that they result from cells that have begun to divide uncontrollably.

Organ affected by cancer	New cases reported each year in the UK (per 1000 people)
Bladder	13.2
Brain	4.0
Bowel (colon and rectum)	34.0
Breast	34.6
Cervix	3.5
Lung	42.2
Lymph system (lymphoma)	9.3
Prostate gland	18.7
Skin (melanoma)	5.0

Table 12.2 The incidence of common cancers in the UK.

Q 5 (a) Which cancer is most common in the UK?

(b) Suggest one reason to explain why this cancer is so common.

(c) Name one type of cancer that is found:

(i) only in women;

(ii) only in men.

Tumours

Mitosis was described in Chapter 10. Normally, mitosis is a controlled process. Cells divide during growth or repair but then stop. Only in a few parts of the human body, such as the upper layer of the skin (epidermis) and the lining of the intestine, do cells keep dividing. If the regulatory processes that control cell division break down, a cell continues to divide and forms a mass of cells. Such a mass of cells is called a **tumour**.

Q 6 Suggest why the on-going division of cells in the epidermis and the lining of the intestine does not normally cause a tumour.

Benign tumours and malignant tumours (cancers)

Figure 12.6
This benign tumour might look unsightly, but is not life-threatening.

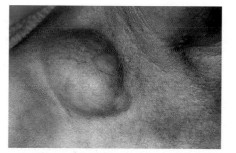

If the mass of cells is harmless, the tumour is said to be **benign**. Although unsightly, the benign tumour shown in Figure 12.6 will not destroy other tissues in the body. At worst, it will press on blood vessels or nerves, interfering with their normal function. **Cancers** are not harmless, they are **malignant tumours**. Cells from malignant tumours:

Figure 12.7
(a) A cancer cell divides repeatedly. (b) The cancer cells invade surrounding tissues, damaging them. (c) If cancer cells break away and enter a blood vessel, they will spread to other parts of the body and form secondary cancers.

- invade other tissues, causing considerable damage;

- break off, spread to other parts of the body via the blood or lymph system and form secondary growths in other organs. This process, called **metastasis**, is shown in Figure 12.7.

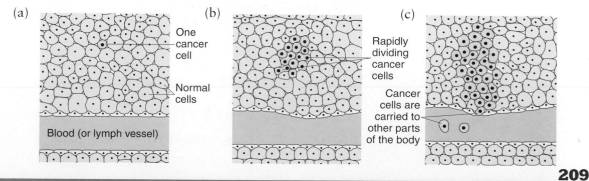

key term

A **tumour** is a mass of cells that continue to divide uncontrollably. Tumours that do not invade other tissues and do not metastasise are called **benign tumours**. Tumours that do invade other tissues and do metastasise are called **malignant tumours** – or cancers.

Figure 12.8
A cervical smear. The blue cells are normal, the red cells are cancer cells.

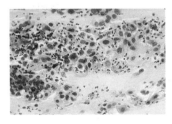

If tumours form inside the body, it is not always obvious whether they are benign or malignant. To find out, a surgeon will remove a small sample of the tumour for analysis. This is called a **biopsy**. In areas prone to cancer and where it is easy to remove cells, routine samples can be taken. In either case, the sample of tissue is stained and examined under an optical microscope. Figure 12.8 shows a cervical smear. The blue cells are normal cells from the cervix . The red cells are cancer cells. Notice how their nuclei take up more space in the cytoplasm than in the normal cells. Can you explain why?

Genes and cancer

If you have forgotten how transcribed genes control protein structure, it might be a good idea to refresh your memory by re-reading the sections on transcription and translation in Chapter 9.

Three types of evidence support the theory that there is a genetic element to cancer.

- The tendency to develop certain types of cancer is inherited.

- In some types of cancer, the tumour cells have characteristic, but abnormal, chromosomes.

- There is a positive correlation between the ability of agents to cause cancer (carcinogens) and their ability to cause mutations (mutagens).

Oncogenes are genes that cause cells to become cancerous. They are formed when the normal version, called a proto-oncogene, mutates so that it becomes overactive.

The normal proto-oncogenes of *ras* oncogenes code for plasma membrane proteins (called G-proteins) that enable cells to respond to growth factors. Normally the G-proteins are inactivated by one of their own enzymes, called GTPase. A cell with a mutant *ras* gene produces G-proteins that are deficient in the enzyme GTPase. As a result, they are active for longer than normal. Almost 30% of human cancers are associated with mutant *ras* oncogenes.

The *myc* proto-oncogene is located on human chromosome 8. The protein encoded by the *myc* proto-oncogene stimulates transcription of genes required for cell division. In a common mutation, this proto-oncogene switches from chromosome 8 to a site on chromosome 14. In its new position, the gene acts as an oncogene and over-stimulates cell division.

The transformation of a normal cell into a fully malignant cell is a multi-step process. It is unlikely that a mutation to produce either the *ras* oncogene or the *myc* oncogene on its own would produce a malignant cell. However, if both oncogenes are present together, full transformation to a malignant cell will result.

Tumour suppressor genes are also associated with cell division. They act in a different way to oncogenes. Whereas proto-oncogenes are converted to oncogenes by mutations that increase the gene's activity, tumour suppressor genes are converted to oncogenes by mutations that reduce their normal activity. A normal, unmutated tumour suppressor gene inhibits cell division.

Retinoblastoma is a childhood tumour of the eye and is an example of a cancer that is caused by the loss of a tumour suppressor gene. In this case the tumour suppressor gene, called *RB1*, is located on human chromosome 13: it inhibits the transcription of proto-oncogenes such as *myc*. With a mutated *RB1*, the *myc* gene becomes overactive and a tumour results. Unexpectedly, *RB1* mutations have also been found in breast, colon and lung tumours.

Another tumour suppressor gene, called *p53*, is located on the short arm of human chromosome 17. It is an important tumour suppressor gene since mutations involving this gene have been associated with nearly 50% of human cancers. Figure 12.9 summarises how this gene is thought to work. Once transcribed and translated, it produces a *p53* protein. Two *p53* proteins join together to form a dimer. If one of the arms of this dimer is an inactive form of *p53*, resulting from a mutation in the normal *p53* gene, the dimer is inactive and tumours result.

Figure 12.9
p53 is a tumour suppressor gene. Two of the protein molecules that it encodes join together to form a dimer. If one or more of these proteins is inactive, resulting from a mutation in the *p53* gene, the dimer will not be effective in suppressing tumours.

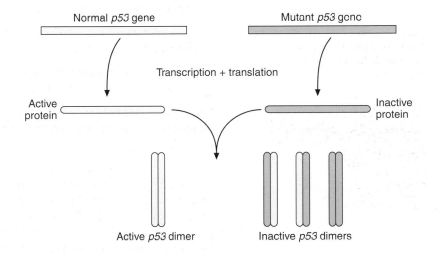

Q 7 (a) In normal human cells, the *p53* protein has a half life of about 6 minutes. Explain what this statement means.

(b) In cancer cells, the *p53* protein has a half life of about 6 hours. What effect will this have on the cell?

Carcinogens

Carcinogens are agents that increase the likelihood of cancer. Some agents are carcinogens because they make mutation of proto-oncogenes or tumour suppressor genes more likely. Common carcinogens fall into three groups.

● Chemical carcinogens, such as ethanol (alcohol).

● Radiation, especially short-wave radiation such as the ultra-violet B radiation found in strong sunlight or the X-rays used in hospitals.

- Viruses, for example the mouse mammary tumour retrovirus (MMTV) causes breast cancer in mice.

Q 8 Within the past two years, scientists have discovered a chemical in cigarette smoke that damages the *p53* gene. What effect will this have on affected cells?

Extension box

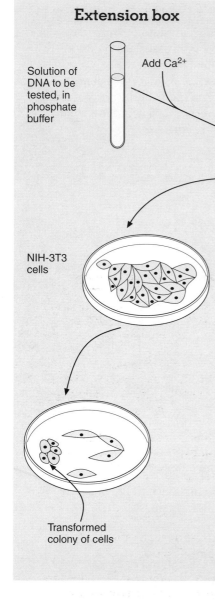

Solution of DNA to be tested, in phosphate buffer

Add Ca²⁺

Co-precipitate of calcium phosphate and DNA

NIH-3T3 cells

Transformed colony of cells

Testing for oncogenes

The study of oncogenes is likely to lead to new methods for controlling cancers. In order to study oncogenes, they must first be isolated. NIH-3T3 cells are fibroblast cells (connective tissue cells that grow well in laboratory cultures) from mice. If injected into immune-deficient mice, these cells do not cause tumours. Laboratory cultures of these cells are used when attempting to isolate oncogenes.

In the test, a solution of extracted DNA is mixed with a solution containing calcium ions. This causes the DNA to form a precipitate. This precipitated DNA is poured on to a culture of NIH-3T3 cells in a Petri dish (Figure 12.10). As a result, a few cells take up the DNA and begin to transcribe and translate it. These cells are said to be transfected cells.

If the DNA in a transfected cell contains oncogenes, the NIH-3T3 cells change their growth pattern. Figure 12.10 shows normal NIH-3T3 cells and a small colony of NIH-3T3 that have become cancerous after being transfected with DNA. This shows that the DNA contained an oncogene.

The advantages of this test are:

- it is technically simple to perform
- NIH-3T3 cells are particularly good at taking up, and expressing, DNA
- cultures of cells are used, rather than whole animals. This removes the need for animal experimentation, which many people oppose, and makes the test suitable for screening large number of DNA samples
- results are obtained fairly quickly.

Figure 12.10

Cigarette smoking and lung cancer

It is estimated that at least 111 000 people in the UK are killed by smoking each year. About 24% of the deaths from CHD in men and 11% of the deaths from CHD in women are due to smoking. In addition, smoking contributes to deaths through lung cancer, bronchitis and emphysema.

In a healthy lung, the movement of mucus, caused by cilia lining the airways, protects the lung from irritants. When a person smokes, the action of the cilia stops and eventually the cilia are destroyed. As a result, tar-laden mucus enters the lungs and is deposited there (Figure 12.11). In addition, the 43 or more carcinogens in tobacco smoke cause mucus-secreting cells to divide uncontrollably. So clear is the relationship between cigarette smoking and premature death that the UK Government issued smoking targets in 1998. Table 12.3 summarises these targets.

Figure 12.11
This lung was removed from a smoker, after his death. The black areas are deposits of tar.

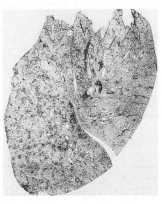

Target group	Targets for England	Targets for Scotland	Targets for Wales
Adults	Reduce smoking rate from 28% of adults to 24% or less by 2010	Reduce smoking from 35% of adults to 31% by 2010	Reduce smoking to no more than 20% of adults by the year 2002
Pregnant women	Reduce smoking rate from 23% of pregnant women to 15% by 2010	Reduce smoking rate from 29% of pregnant women to 20% by 2010	To increase the number of women who give up smoking in early pregnancy to at least 33%
Children	Reduce smoking from 13% of children to 9% or less by 2010	Reduce smoking from 13% of 12–15 year-olds to 11% by 2010	Reduce smoking to 16% of 15 year-old boys and 20% of 15 year-old girls

Table 12.3 Smoking targets for the UK.

Q 9 Suggest why the smoking targets, shown in Table 12.3, are different in England, Scotland and Wales.

Although Government targets and health education programmes are clearly useful in focusing on the damage to our health caused by smoking, some people do not think they go far enough. Look at Figure 12.12. Although the man is not smoking, he is inevitably breathing the cigarette fumes of the woman who is smoking. This passive smoking also damages the health of the non-smokers.

Many organisations have introduced 'no smoking policies', preventing people from smoking in their buildings. This helps cut down passive smoking as well as reducing the frequency of smoking amongst smokers. Some people believe that more drastic measures should be introduced, however. These include banning smoking from all public places, introducing even higher taxes on the sale of tobacco, banning advertisements for smoking and even banning the sale of tobacco.

Figure 12.12
The non-smoker in this picture suffers the harmful effects of the smoke from the cigarette smoker. Many public buildings have 'no-smoking policies' to reduce the effects of passive smoking.

Against such suggestions, many smokers' organisations argue that people should be free to make their own choice about whether or not to smoke. They argue that personal freedom is at stake and reject the notion of a 'big brother' establishment that restricts this personal freedom. They also point out that the sale of cigarettes in the UK has fallen from 96 748 million in 1987/88 to 91 675 million in 1997/98, so that existing measures are having an effect without the need for further restrictions.

Melanoma

Human skin protects the body against water loss, infection and radiation. It is made of two layers, the upper epidermis and the lower dermis. Cells called melanocytes are found in the epidermis layer and produce **melanin**, which gives the skin its colour. **Melanoma** is a disease of the skin in which cancer cells develop in the layer of melanocytes.

Melanoma is extremely dangerous since it can metastasise quickly to other parts of the body through the blood or lymph systems. It is usually detected when a skin mole changes in shape or size, starts to bleed or becomes itchy. If any of these symptoms appears, an immediate visit to a doctor is necessary.

Figure 12.13
These holiday makers are enjoying sunbathing but, being fair-skinned, are risking damage to cells in their skin that can give rise to melanoma.

The most common cause of melanoma is exposure to radiation. The most important source of harmful radiation is ultra-violet light in sunlight, especially UVB rays that damage DNA. To some extent, melanin filters the harmful radiation, so protecting the skin. Fair-skinned people have little melanin in their skins, however, and are prone to melanoma if they expose themselves to the sun (Figure 12.13). Men most often get melanoma on the trunk of their bodies (the area between the shoulders and hips) whereas women most often get melanoma on the arms and legs.

Q 10 Suggest why melanoma is more common amongst fair-skinned people living in Australia than those living in England.

Summary

- Coronary heart disease CHD occurs when fatty materials (atheromas) build up within artery walls, eventually blocking the artery. Blockage of a coronary artery results in the death of part of the heart muscle (myocardial infarction, or heart attack).

- The presence of atheromas increases the risk of arteries bursting (aneurysms) and blood clots forming within arteries (thrombosis).

- Many factors increase the risk of CHD. Some, such as your age or inheritance, cannot be avoided. Others, such as diet, smoking cigarettes or high blood pressure, can be changed.

- Low-density lipoproteins contain high levels of cholesterol and increase the risk of developing atheroma. High-density lipoproteins have low cholesterol levels and are less dangerous to health.

- Tumour cells are cells in which the control mechanisms regulating cell division have broken down. As a result they divide uncontrollably.

- Benign tumour cells do not invade other tissues and destroy them.

- Malignant tumours (cancers) invade and damage other tissues. Malignant cells metastasise by breaking off from tumours and invading other parts of the body via the blood or lymph systems.

- Oncogenes are genes that cause cells to become cancerous. They are formed by the mutation of a proto-oncogene or of a tumour suppressor gene.

- Chemical carcinogens and radiation may damage DNA and cause the production of oncogenes.

- Lung cancer is caused by carcinogens in tobacco smoke. It is advisable not to smoke.

- Melanoma is one type of skin cancer affecting melanin-producing cells in the skin. Exposure to the ultra-violet light in sunlight is one factor that causes the formation of melanomas.

Assignment

Finding a cure

In the assignment at the end of Chapter 8, we looked at some work that was carried out over 200 years ago. In this assignment we will look at some much more modern work concerned with treating cancer.

As you saw earlier in this chapter, one of the big problems with cancer is that it isn't a single disease. There are lots of different types of cancer. None of the research described in this assignment is going to provide an instant cure. But . . . these ideas may lead to more successful ways of treating some types of cancer and that will be a step forward.

Let's remind ourselves of what we know about cancer. Study Figure 12.14. Look first at the diagram and the information in the blue boxes. It summarises some things about cancer cells and how they spread to different parts of the body and produce secondary tumours.

Now look at the red boxes. Most of them contain information about new research. A lot of this research is very complicated and difficult to understand. Biologists need to be aware of developments in their subject including those outside their own fields of study. They use their knowledge to get an understanding of the basic ideas even if they do not follow all the detail. So, see if you can use your knowledge of biology to answer the questions in this assignment. When you have finished this assignment you should have a good understanding of some of the discoveries that might, one day, lead to more effective ways of controlling cancer.

Now answer the questions. The question numbers correspond to the numbers used for the red boxes. You will need to look at the information in the relevant box before you answer the question.

1. (a) Chemotherapy means using drugs for killing cancer cells. Explain why drugs which destroy DNA will kill cancer cells.

    (1 mark)

    (b) Drugs which destroy DNA often have unpleasant side effects because they kill any cells that are dividing rapidly. One side effect is that the patient's hair often falls out. Suggest why. *(2 marks)*

2. (a) Use your knowledge of cell organelles to explain why destroying its mitochondria will kill a cell. *(2 marks)*

    (b) Explain why it is thought that CT2548 will kill cancer cells but will not kill healthy cells. *(2 marks)*

3. (a) The functions of many proteins depends on their tertiary structure. Explain why all the molecules of a particular sort of protein always have the same tertiary structure. *(2 marks)*

    (b) Explain, in terms of the shape of protein molecules:
    (i) why the antibodies injected when using ADEPT only target cancer cells;
    (ii) how the inactive pro-drug is converted into its poisonous active form. *(4 marks)*

Figure 12.14

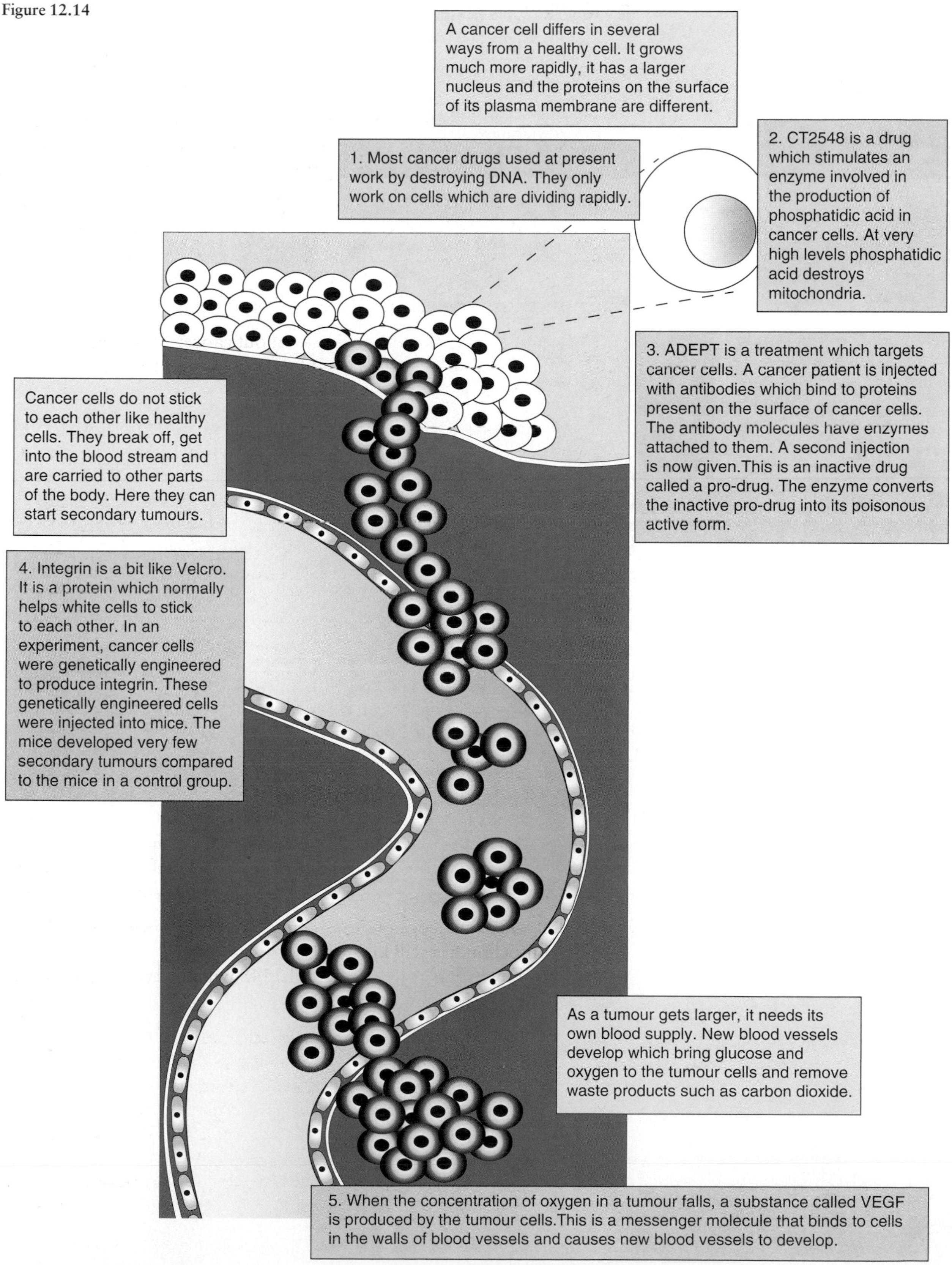

A cancer cell differs in several ways from a healthy cell. It grows much more rapidly, it has a larger nucleus and the proteins on the surface of its plasma membrane are different.

1. Most cancer drugs used at present work by destroying DNA. They only work on cells which are dividing rapidly.

2. CT2548 is a drug which stimulates an enzyme involved in the production of phosphatidic acid in cancer cells. At very high levels phosphatidic acid destroys mitochondria.

3. ADEPT is a treatment which targets cancer cells. A cancer patient is injected with antibodies which bind to proteins present on the surface of cancer cells. The antibody molecules have enzymes attached to them. A second injection is now given. This is an inactive drug called a pro-drug. The enzyme converts the inactive pro-drug into its poisonous active form.

Cancer cells do not stick to each other like healthy cells. They break off, get into the blood stream and are carried to other parts of the body. Here they can start secondary tumours.

4. Integrin is a bit like Velcro. It is a protein which normally helps white cells to stick to each other. In an experiment, cancer cells were genetically engineered to produce integrin. These genetically engineered cells were injected into mice. The mice developed very few secondary tumours compared to the mice in a control group.

As a tumour gets larger, it needs its own blood supply. New blood vessels develop which bring glucose and oxygen to the tumour cells and remove waste products such as carbon dioxide.

5. When the concentration of oxygen in a tumour falls, a substance called VEGF is produced by the tumour cells. This is a messenger molecule that binds to cells in the walls of blood vessels and causes new blood vessels to develop.

4. (a) Describe, in outline, the main steps involved in genetically
engineering cancer cells to produce integrin. *(4 marks)*
   (b) Explain why integrin could only be used as a cure for cancer when
used with other drugs. *(1 mark)*
   (c) In the experiment described, why was it necessary to have a
control group of mice. *(2 marks)*

5. A search is being made for substances which will stop VEGF
working. Suggest how these substances could:
   (a) lead to the death of cancer cells; *(2 marks)*
   (b) have very little effect on other cells in the body of an adult. *(2 marks)*

Examination question

Table 12.4 shows the percentage of certain groups of people who smoked
cigarettes.

Age (years)	Sex	1972 (%)	1974 (%)	1976 (%)	1978 (%)	1980 (%)	1982 (%)	1984 (%)	1986 (%)	1988 (%)	1990 (%)	1992 (%)	1994 (%)
16–19	Female	39	38	34	33	32	30	32	30	28	32	25	27
	Male	43	42	39	35	32	31	29	30	28	28	29	28
20–24	Female	48	44	45	43	40	40	36	38	37	39	37	38
	Male	55	52	47	45	44	41	40	41	37	38	39	40

Table 12.4

(a) What does the table show about the pattern of cigarette smoking:
   (i) of different age groups;
   (ii) of females and males. *(2 marks)*
(b) Suggest **two** explanations for the change in cigarette smoking
between 1972 and 1994. *(2 marks)*
(c) Describe **two** ways in which cigarette smoking can damage the health
of smokers. *(2 marks)*

13 Diagnosing and Controlling Disease

Figure 13.1 shows cells of a bacterium called *Staphylococcus aureus*. Most people carry *S. aureus* in their noses, mouths and on their skin, where it usually does no harm. Occasionally, *S. aureus* causes infections, such as boils, abscesses and conjunctivitis.

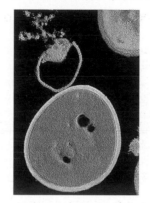

Figure 13.1
Cells of the bacterium *Staphylococcus aureus*. The smaller cell is bursting, as a result of treatment with antibiotic.

Normally, the skin is an effective barrier to *S. aureus*. When the skin is breached by a wound or by surgery, *S. aureus* can enter the body creating an abscess lesion. This consists of dead tissue, fibrin (a clotting agent in blood) and a large number of living and dead white blood cells. The toxin released by this infection can affect other parts of the body. At any time, the multiplying bacteria can break away and travel through the body in the blood or lymphatic systems. This results in septicaemia, which can lead to secondary infections and eventually death.

Open-body surgery is only possible because patients can be protected from infection by bacteria such as *S. aureus*. Initially, the antibiotic penicillin worked against *S. aureus*. However, by the 1950s, *S. aureus* had become resistant to penicillin and all other antibiotics that were known at that time. This made open-body surgery a risky business once again.

As a result of extensive research by pharmacological companies, a new generation of semi-synthetic penicillins were developed. These included methicillin, oxacillin, cloxacillin and nafacillin. They became best-selling drugs, specifically because they worked against penicillin-resistant *S. aureus*. However, *S. aureus* fought back and, by 1976, a strain resistant to methicillin and gentamycin was reported in the UK, in Ireland, in Australia and in the USA. This methicillin-resistant *S. aureus* (MRSA) caused havoc in hospitals throughout the world.

MRSA can make common operations, like hip replacements, very hazardous. It can also cause serious infections in people with intravenous catheters and prosthetic heart valves. Hospital patients with weakened immunity are also at risk. These include premature babies, people who have suffered major burns, and intensive-care patients. Hospitals in the UK now have isolation wards and specially trained nurses to deal with infections of MRSA, and other 'superbugs' developed by in-patients.

People suffering a disease usually notice its effects on their body. These effects are called **symptoms**. In addition, there are other effects that the sufferer might not notice. These will be picked up by a doctor and are called **signs**. A doctor uses clinical signs to help in diagnosing a disease. In this chapter, we will examine some of the techniques used in disease diagnosis. The first involves DNA. If you have forgotten the structure of DNA, you should revise it by re-reading Chapter 9. You will also find aspects of genetic engineering, which are described in Chapter 11, are referred to again in this chapter. Later in the chapter, we will learn about some of the drugs that are used to treat diseases, including antibiotics, such as penicillin.

Using DNA probes to diagnose disease

In Chapter 9, we learned how diseases such as cystic fibrosis and phenylketonuria are caused by defective genes. We have also learned in Chapter 12 that some cancers are caused by oncogenes. Since human genes are made of DNA, we can use DNA probes to identify genes associated with disease. This technique will only work if we know some of the specific base sequences found in the defective gene we wish to identify in human chromosomes. Once we know this sequence, we can make single-stranded complementary DNA (cDNA) that matches this base sequence (see Chapter 11).

DNA extraction

To find disease-causing genes, we must first obtain a sample of DNA. We can obtain this sample from most human cells. In one technique, a small sample of tissue (e.g. 0.5 cm³ of blood) is shaken in the laboratory with a mixture of water-saturated phenol and chloroform. This causes proteins to precipitate out, leaving DNA in the water layer. This DNA is then extracted from the water layer and purified.

Q 1 Use your knowledge from Chapter 6 to identify which cells in a blood sample contain DNA.

DNA fragmentation

The DNA in our sample is then cut into fragments. You have seen in Chapter 11 that restriction endonucleases (**restriction enzymes**) are enzymes that cut DNA at specific base sequences, called recognition sequences. These enzymes are used to cut DNA in our sample, producing a complex mixture of DNA fragments of different lengths (Figure 13.2).

Figure 13.2
Extracted DNA is cut into fragments of different sizes using restriction enzymes.

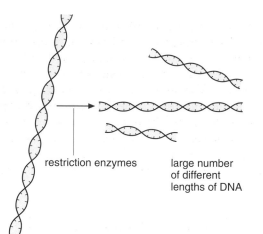

restriction enzymes

large number of different lengths of DNA

key terms

Restriction enzymes
cut DNA into fragments.

Electrophoresis
separates DNA fragments according to their size and electrical charge.

Separation of DNA fragments

Electrophoresis is used to separate the DNA fragments according to their electrical charge and their size. This involves placing the DNA mixture in a well at one end of a block of agarose gel (Figure 13.3). When an electrical current is passed through the gel, the negatively charged DNA fragments move to the positively charged electrode. The smaller fragments move more quickly through the gel than the larger fragments. As a result, they travel further through the gel. This process separates the different DNA fragments into bands.

Q **2** Use your knowledge from Chapter 9 to explain why DNA fragments have a negative charge.

Location of specific DNA sequences using a DNA probe

Immediately after electrophoresis, the DNA bands are not visible. To make them visible we must add a DNA probe. A DNA probe is a single strand of complementary DNA (cDNA), which is labelled with a radioactive isotope. It can attach to other single-stranded DNA of the matching base sequence by base pairing (see Chapter 9). However, our DNA fragments are double stranded. We must separate the double-stranded DNA fragments on our agarose gel into single strands. This is done by immersing the gel in an alkaline solution (Figure 13.4).

Figure 13.3
The DNA mixture is placed in a well at one end of a gel. When an electric current is applied, the DNA fragments move to the positively charged electrode, according to their size. This is the process of electrophoresis.

Separation of fragments according to size and electrical charge

Figure 13.4
Double-stranded DNA is separated into single-stranded DNA.

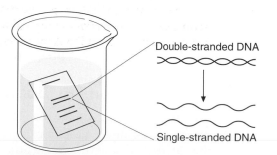

Gel placed in a solution to separate double-stranded DNA

Double-stranded DNA

Single-stranded DNA

A **DNA probe** is a fragment of single-stranded DNA that has the complementary base sequence to part of a gene we are looking for. It is radioactively labelled.

The single-stranded fragments of DNA are now transferred from the gel to a nylon membrane. This is done by placing a nylon membrane over the gel and covering it with absorbent paper towels. The absorbent paper draws the DNA fragments from the gel into the membrane by capillary action. When the membrane is removed from the gel (Figure 13.5), the DNA fragments are located on the membrane in exactly the same position as on the gel. These fragments can be fixed to the membrane by exposure to UV light.

Figure 13.5
When the nylon membrane is removed from the gel, it contains the DNA bands in exactly the same position as they were on the gel.

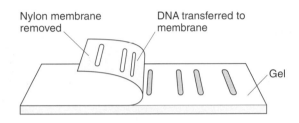

We can now locate the DNA sequences we are looking for. The nylon membrane is immersed in a solution containing the radioactive DNA probe. If the matching base sequence of our disease-causing gene is present on any of the DNA fragments, the DNA probe attaches to it. This makes these fragments radioactive (Figure 13.6).

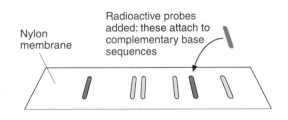

Figure 13.6
Radioactive probes bind to target sequences of DNA, making them radioactive.

The radioactive solution is then washed off the membrane. In a darkroom, an X-ray plate is placed over the membrane, which is left for several hours. The places where radioactive probes have bound to DNA fragments will give off radiation, causing the X-ray film to 'fog'. Once developed, this causes a pattern of bands on the X-ray plate, confirming the presence of our disease-causing gene.

Q 3 How does a DNA probe attach to DNA fragments?

Using enzymes to diagnose disease

The chemical balance of the blood often changes during the course of an illness. For example, if cells are damaged they release their contents into the blood. As damaged cells are broken down, products of their digestion appear in the blood. By detecting the presence of these chemicals, we can often diagnose particular diseases.

Detecting the body's own enzymes to diagnose pancreatitis

The pancreas is an organ that lies just under the stomach (Figure 13.7). It produces hormones that regulate blood sugar levels. Insulin-dependent diabetics have lost the ability to secrete one of the hormones, insulin, from cells in their pancreas. The pancreas also secretes a number of digestive enzymes into the gut. These enzymes include amylase, which breaks down starch, lipase, which breaks down lipids (fats), and trypsin, which breaks down proteins.

Figure 13.7
The pancreas lies just underneath the stomach.

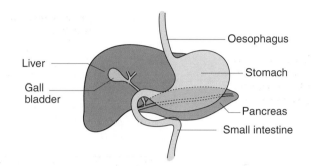

Pancreatitis is a disease involving severe inflammation of the pancreas. During this inflammation, pancreatic cells become damaged and release their enzymes abnormally. Testing for the presence of these enzymes helps in the diagnosis of pancreatitis.

- Acute pancreatitis occurs suddenly. It is diagnosed by the presence of pancreatic enzymes, principally amylase and lipase, in the blood of sufferers.

- Chronic pancreatitis is a long-term condition in which the pancreas gradually loses its ability to produce enzymes. This type of pancreatitis is diagnosed by testing for an unusually low level of pancreatic enzymes in the faeces of a sufferer.

Q 4   Suggest why:

    (a)   pancreatic enzymes are normally found in human faeces;
    (b)   unusually low levels of pancreatic enzymes are found in the faeces of people suffering chronic pancreatitis.

Using enzymes in biosensors

You learned in Chapter 3 how to use Benedict's reagent to detect reducing sugars, such as glucose. For diabetics, who must monitor their own blood glucose levels, the Benedict's test is inconvenient. A biosensor has been devised that provides an accurate and quick alternative.

The biosensor relies on two enzymes: glucose oxidase and peroxidase. Glucose oxidase is particularly important in this process because:

- it is highly sensitive – enabling very low concentrations of glucose to be detected

- it is highly specific – reacting only with glucose.

Glucose oxidase catalyses the conversion of glucose to hydrogen peroxide (H_2O_2), shown by the equation below.

$$\text{glucose} + O_2 \xrightarrow{\text{glucose oxidase}} \text{gluconic acid} + H_2O_2$$

This reaction is coupled to a second reaction involving hydrogen peroxide and a hydrogen-donor molecule (represented as **D** in the equation below). This reaction is catalysed by peroxidase.

$$\underset{\text{(colourless)}}{DH_2} + H_2O_2 \xrightarrow{\text{peroxidase enzyme}} 2\,H_2O + \underset{\text{(coloured)}}{D}$$

Molecules of the two enzymes and of the colourless hydrogen donor are immobilised on an inert strip. This strip is dipped into a test solution, such as urine. If the test solution contains glucose, a coloured compound appears on the inert strip. The colour that develops indicates the concentration of glucose in the test solution. This is the basis of the Clinistix™ test (Figure 13.8), used by diabetics to monitor their blood glucose levels.

Figure 13.8
A Clinistix™ test is used to test for glucose in urine.

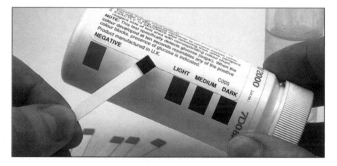

Q 5 Suggest why it is more convenient for a diabetic to test a urine sample for glucose with Clinistix™ rather than with Benedict's solution.

The use of drugs in the treatment and control of disease

A wide range of drugs is available to control diseases. Here we will examine three groups of drugs:

● beta blockers – used to control hypertension

● antibiotics – used to control bacterial infections

● monoclonal antibodies – used to ensure that treatments target specific cells.

Beta blockers

Some people suffer prolonged, high blood pressure. This is called **hypertension.** As we saw in Chapter 12, hypertension is a risk factor increasing the likelihood of coronary heart disease.

Just as skeletal muscle increases in size in response to exercise, the smooth muscle in artery walls increases in size in a person suffering hypertension. This increase in size makes the lumen of the artery smaller, which further raises the blood pressure. Figure 13.9 shows the links between hypertension and an increased risk of cardiovascular diseases. The heart of someone suffering from hypertension also contracts more frequently and with greater force than usual.

Figure 13.9
The links between hypertension and cardiovascular disease.

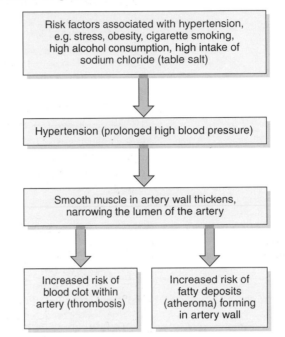

We learned in Chapter 2 that plasma membranes are mainly lipid, but contain many proteins. Some of these proteins 'recognise' specific molecules and enable cells to respond to them. These membrane proteins, called receptors, are able to 'recognise' chemical messengers, such as hormones, because they have a complex three-dimensional shape that complements the shape of the chemical messenger. In this way, they act in a similar way to the lock-and-key model of enzyme action that you learned about in Chapter 4. The normal chemical messengers that fit into, and stimulate, receptors in plasma membranes are called **agonists.**

Beta (β) receptors are found on the plasma membranes of muscle cells in the wall of arteries and arterioles (small arteries) and in the heart. The chemical messengers that fit into, and stimulate, β receptors are called β-**agonists.** Once their β receptors are stimulated, muscle cells in the walls of arteries relax. This makes the lumen of each affected artery bigger.

Some molecules can mimic the action of agonists. Because they have a similar shape to the agonists, they can bind to the receptors on plasma membranes, rather like competitive inhibitors can bind to the active site of an enzyme (see Chapter 4). Molecules that mimic agonists are called **antagonists.** They are also called blockers because they block receptor

key terms

Agonists
are chemicals that stimulate specific receptors on plasma membranes.

Antagonists
are chemicals that mimic the action of natural agonists.

β blockers
are antagonists of β-agonists and so block β receptors on the surface of muscle cells in arteries and in the heart.

sites, preventing the normal agonist from binding to it. β **blockers** mimic the action of β-agonists. In this way, they reduce hypertension (see Figure 13.10). Propanolol, the drug named in Figure 13.10, is non-selective. A second β blocker, called Atenolol, is selective, affecting only cardiac muscle cells.

Figure 13.10
β blockers reduce hypertension.

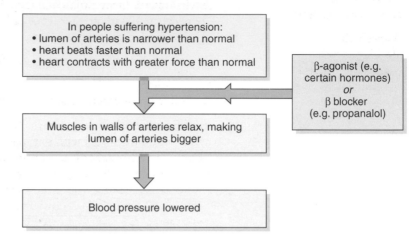

Q 6 A diuretic is a chemical that increases the rate of urine production. Suggest why diuretics are often included in the treatment of people taking β blockers.

Antibiotics

An antibiotic is a chemical, produced by a microorganism that, in dilute concentrations, harms or kills another bacterium. (You should note that antibiotics do not harm viruses. The reason a GP will often prescribe an antibiotic when you have a viral infection is to protect you from any bacterial infection that might get a hold whilst your immune system is dealing with the virus.) The great majority of antibiotics are produced as natural secretions by bacterial or fungal cells. About 100 antibiotics are commonly in use today, of which 90% are isolated from bacteria. There are a few synthetic antibiotics, i.e. antibiotics that have been made from scratch in the laboratory.

Q 7 Do antibiotics harm human cells? Explain your answer.

Figure 13.11
Microorganisms produce antibiotics late in their life cycle.

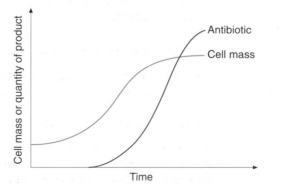

key terms

Bacteriostatic antibiotics slow down the growth of pathogenic bacteria.

Bactericidal antibiotics kill pathogenic bacteria.

Figure 13.11 shows that bacteria and fungi produce antibiotics during a late stage of their life cycle. For this reason, they are called secondary metabolites. The value of an antibiotic to an antibiotic-secreting cell seems to be that it inhibits the growth of natural competitors, giving the antibiotic-secreting population an advantage in colonising its environment. Once isolated and purified, antibiotics are useful to us because they harm disease-causing bacteria (**pathogenic bacteria**). They can harm pathogenic bacteria either by:

- slowing down their growth rate. **Bacteriostatic** antibiotics work in this way

- killing them. **Bactericidal** antibiotics work in this way.

Antibiotics work in many ways. Table 13.1 summarises three ways in which antibiotics deal with pathogenic bacteria. Note that in a Unit Test, you will not be expected to link specific antibiotics with their mode of action. Some antibiotics, such as penicillin, are effective on only a few pathogenic bacteria. They are called **narrow-spectrum antibiotics**. Other antibiotics, such as chloramphenicol, are effective against many types of pathogenic bacteria. They are called **wide-spectrum antibiotics.**

Action of antibiotic	Explanation of effect of antibiotic
Prevents formation of cell walls by bacterial cells.	The cell wall stops bacterial cells swelling and bursting in solutions with a more negative water potential than their own cytoplasm (see Chapter 2). Without their cell walls, affected bacterial cells swell and burst in human plasma.
Prevents DNA transcription in bacterial cells (see Chapter 9 for an account of DNA transcription – the process by which mRNA is made from a DNA gene).	Affected cells are unable to transcribe essential genes and so die.
Prevents mRNA translation in bacterial cells (see Chapter 9 for an account of translation – the process by which mRNA is used to make proteins).	Affected cells are unable to make essential proteins and die.

Table 13.1 Three different ways in which antibiotics affect pathogenic bacteria.

Q 8 Distinguish between a bacteriostatic antibiotic and a bactericidal antibiotic.

Monoclonal antibodies

When we suffer an infectious disease, we usually recover from it. This recovery depends on a group of proteins, called **antibodies**. In humans, these substances are secreted by white blood cells, called **B-lymphocytes**.

The antibodies are produced in response to the presence of a foreign chemical that has entered the body, called an **antigen.** All human cells have antigens on their plasma membranes. These are called **self-antigens** and we do not normally produce antibodies against them. Antigens on the surface of foreign cells (**non-self-antigens**) are different from self-antigens (Figure 13.12): they cause B-lymphocytes to produce antibodies.

Figure 13.12
Antigens present on the cells of your own body (self-antigens) are different from those present on foreign cells (non-self-antigens).

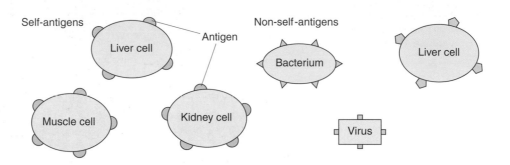

Each antibody is highly specific. It is produced by only one type of B-lymphocyte in response to only one type of antigen. Figure 13.13 represents three different B-lymphocytes and the antibodies that each makes. Each antibody is a Y-shaped molecule with specific, antigen-binding sites at the ends of its 'arms'. Note that these binding sites are identical to the receptor sites on the plasma membranes of B-lymphocytes that make the antibodies and exactly fit only one type of antigen.

Figure 13.13
B-lymphocytes are present in a variety of forms, each having the ability to produce one specific type of antibody.

B-lymphocytes with different specific receptors

Antibody produced by each different B-lymphocyte

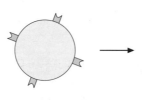

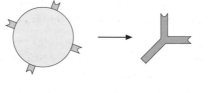

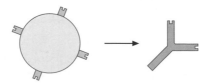

key terms

An antigen
is a chemical found on the surface of a cell.

Self-antigens
are found on cells in your own body.

Non-self-antigens
are found on cells of other organisms, including other humans.

Figure 13.14 shows what happens when a specific antigen is detected in the body. A specific B-lymphocyte has a receptor site that matches the non-self-antigen. This cell divides rapidly to form a clone of identical cells.

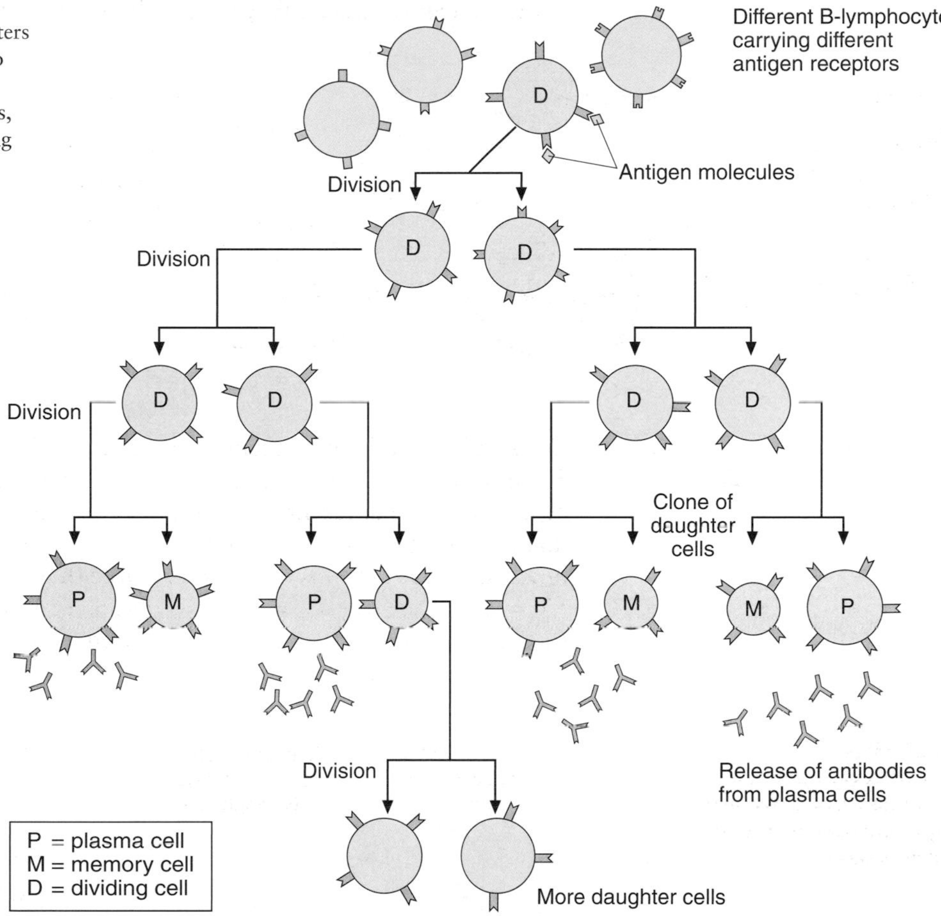

Figure 13.14
When a B-lymphocyte encounters its antigen, it divides rapidly to form a clone of identical cells. Some of these are memory cells, but most are antibody-secreting plasma cells.

Different B-lymphocytes carrying different antigen receptors

Antigen molecules

Division

Division

Division

Division

Clone of daughter cells

Release of antibodies from plasma cells

Division

More daughter cells

P = plasma cell
M = memory cell
D = dividing cell

- Most of these clone cells are **plasma cells,** which release thousands of molecules of their antibodies. The antibodies combine with the non-self-antigen, making it harmless.

- A few clone cells remain as **memory cells,** ready to make antibodies if the non-self-antigen ever enters the body again.

In the laboratory, B-lymphocytes can be fused with tumour cells. The resulting cell is called a **hybridoma**. It inherits two useful properties, one from each parent cell.

- A hybridoma divides rapidly in laboratory cultures to form a clone of identical cells.

- A hybridoma continuously produces specific antibodies. Because the antibodies are of a single type, produced by a single clone of cells, they are called **monoclonal antibodies.**

Q 9 Why do we produce antibodies to a non-self-antigen much faster on second exposure than on first exposure?

Drugs and radiation that are used to treat cancer often affect non-cancerous cells. This causes harmful side effects during treatment. Human cancer cells have surface antigens that are identical to non-cancerous cells. However, some types of cancer cell also have surface antigens that are not found on non-cancerous cells: they are called **tumour markers.** This is where monoclonal antibodies are useful. We can make monoclonal antibodies against these tumour markers. As a result, these antibodies will attach only to cancer cells. If we now attach molecules of an anti-cancer drug to the monoclonal antibody molecules, they will deliver the drug directly to cancer cells. Other, non-cancerous cells will be unaffected. Because this treatment is highly specific, anti-cancer drugs attached to monoclonal antibodies are often called '**magic bullets**'.

Monoclonal antibodies can also be used in the **antibody direct enzyme prodrug therapy** (ADEPT) technique. In this technique, monoclonal antibody molecules that are specific to tumour markers are tagged with an enzyme that converts an inactive drug (the **prodrug**) to an active form that kills cells (i.e. is cytotoxic). After injection, these tagged antibodies attach only to cancer cells. The prodrug is then injected in a relatively high concentration. Being inactive, it has no effect on normal cells but, when it encounters a cancer cell that is attached to a tagged antibody, the attached enzyme activates the drug killing the cancer cells.

key term

A **hybridoma** is a cell that has been formed in the laboratory by the fusion of a B-lymphocyte with a tumour cell. It divides rapidly in the laboratory and produces antibodies of only one type, called **monoclonal antibodies.**

Extension box

Uses of monoclonal antibodies

As well as enabling us to target the treatment of cancer cells, monoclonal antibodies have a range of other uses.

Monoclonal antibodies can be used to screen donated blood for harmful viruses, such as those that cause AIDS or hepatitis. In this way, contaminated blood is detected and rejected for blood transfusion.

In **immunoassays,** monoclonal antibodies are labelled in some way, e.g. radioactively or with a fluorescent dye. This makes them easy to detect. When added to a test sample, many antibodies will attach to their specific antigen. Washing the treated sample in solutions that remove unattached antibodies leaves only those that are attached to their antigen. The quantity of attached antibodies can then be judged by measuring the extent of radioactivity or of fluorescence in the test sample.

In the **enzyme-linked immunosorbant assay** (ELISA) technique, monoclonal antibodies are immobilised on an inert base and test solutions are passed over them. If molecules of the target antigen are present in the test solution, they will combine with the immobilised monoclonal antibodies. A second type of antibody, which has an enzyme attached to its molecules, is then added. It combines only with those immobilised antibodies that are linked to the target antigen. Finally, a substrate is added which is converted to a coloured product by the enzyme attached to the second type of antibody. The amount of colour tells us how much of the antigen was present in the test solution. The ELISA test is used to detect drugs in the urine of athletes. It is also used in home pregnancy tests, like the one shown in Figure 13.15. In home pregnancy tests, the antigen in question is human chorionic gonadotrophin (hCG) that is secreted by the placenta.

Although the immune response described in this chapter helps us to combat disease, there are times when the immune response is not helpful. One of these is during transplant surgery. Cells of transplanted organs have non-self-antigens. These trigger the production of antibodies that attack the transplanted organ, leading to its rejection. Cells called T-lymphocytes are needed for B-lymphocytes to function. Monoclonal antibodies against T-lymphocytes can be used to prevent B-lymphocytes from functioning, thus blocking the rejection of transplanted organs.

Figure 13.15
(a) A home pregnancy testing kit.
(b) A diagrammatic section through a pregnancy testing kit, showing the use of immobilised enzymes.

(a)

(b)

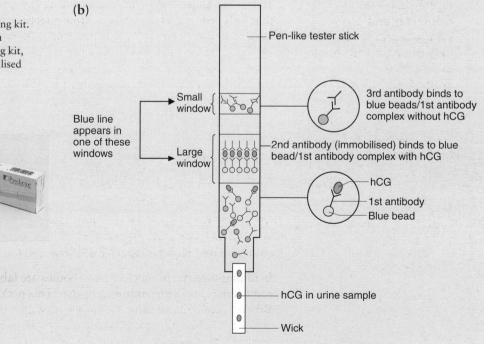

Q 10 Monoclonal antibodies can be used to detect enzymes in the blood that are released by damaged heart muscle cells. This helps to diagnose that someone has suffered a heart attack. Suggest how this technique could be used.

Summary

- A DNA probe is a fragment of single-stranded DNA that has the complementary base sequence to part of a gene we are looking for. It is radioactively labelled.

- DNA probes can be used to identify the presence of a disease-causing gene in extracted DNA.

- Pancreatitis can be detected by the presence of pancreatic enzymes in the blood and by low concentrations of pancreatic enzymes in faeces.

- Because glucose oxidase is specific and highly sensitive, it can be used in conjunction with peroxidase to test for glucose in urine.

- Beta blockers (β blockers) bind to receptors on the plasma membranes of muscle cells in the arteries and in the heart. They can be used to treat hypertension.

- Antibiotics are substances produced by bacteria and fungi that, in dilute concentrations, inhibit the growth of other bacteria. For this reason, they can be used to treat bacterial infections.

- Bacteriostatic antibiotics slow the growth of bacteria whereas bactericidal antibiotics kill bacteria.

- Antibiotics work in a number of ways. These include prevention of DNA replication, prevention of protein synthesis and prevention of bacterial cell wall formation, leading to the osmotic lysis of affected cells.

- Hybridoma cells are made by fusing lymphocytes and tumour cells. They grow rapidly in the laboratory, producing a single type of antibody, called a monoclonal antibody.

- Monoclonal antibodies can be used to target treatment to specific cells and for locating specific chemicals.

Assignment

Birds of a feather

In this chapter you have studied one way in which we make use of our knowledge of the structure of DNA. DNA probes, however, are not only used in diagnosing disease. In this assignment we will look at how they have been adapted to help with the conservation of birds of prey.

Very strict laws apply if you want to keep a hawk or a falcon in captivity. Birds can be imported. A licence is required and this makes it easy to check up to see if a particular bird has been imported. As it is difficult to get a licence and importing birds is expensive, it might be thought better to raise the young produced by captive birds. Unfortunately this is not very easy, so people are sometimes tempted to take the young out of the nests of wild birds. This is illegal as many birds of prey are rare and protected by law. The big problem for those involved in conservation is to prove whether a bird was bred in captivity or taken from the wild. This is where genetic fingerprinting can help.

Figure 13.16
Taking the eggs or young from nests of birds of prey, such as this falcon, is illegal. Genetic fingerprinting enables us to prove whether a particular bird was bred in captivity or taken from the wild.

Let us look at an actual case. A Wiltshire man had a collection of birds of prey which included five peregrine falcons: an adult male, an adult female and three young birds. He claimed that the young birds were the offspring of the adult pair. The police, however, were suspicious and were of the opinion that some or all of these young birds had been taken from the wild.

The technique used to investigate this case involved the use of **single-locus probes**. Look at Figure 13.17. This is a diagram of a single pair of chromosomes from a peregrine falcon. Arranged along each chromosome are the genes which code for proteins. Each gene is found at a particular place or **locus** on the chromosome. Individual genes may exist in different forms called **alleles**. The alleles of a gene have slight differences in their base sequences and it is these variations which give rise to the inherited differences between individuals. The DNA between the genes is sometimes called non-coding DNA. Some of this DNA is made up of short, repeated sequences of nucleotides. The non-coding DNA found at a particular locus of one chromosome of a pair may differ from that found at the same locus on the other chromosome in the number of times the repeated sequence occurs.

Figure 13.17
Diagram representing a single pair of chromosomes from a peregrine falcon.

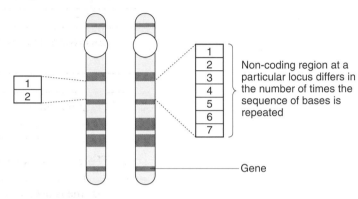

1 Copy figure 13.17. Use the information in the paragraph above to annotate your diagram to show the main features of genes and non-coding DNA.

(2 marks)

A parent will pass one of each pair of its chromosomes to each of its offspring, so the non-coding sequences of DNA will be inherited in the same way that genes are inherited.

2 A female bird has a sequence of 6 repeats at a particular locus on one of its chromosomes. At the corresponding locus on the other chromosome of the pair, there is a sequence of 4 repeats. Explaining your answer in each case, answer the following questions.

(a) What proportion of the offspring of this bird would you expect to have the non-coding piece of DNA with 6 repeats?

(2 marks)

(b) A young bird has a piece of DNA with 4 repeats and a piece of DNA with 3 repeats. Could the female bird have been its mother?

(2 marks)

The main steps in carrying out this method of genetic fingerprinting are shown in the flow chart.

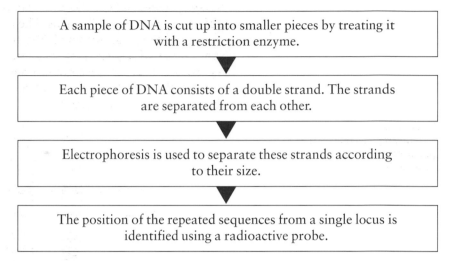

A sample of DNA is cut up into smaller pieces by treating it with a restriction enzyme.

▼

Each piece of DNA consists of a double strand. The strands are separated from each other.

▼

Electrophoresis is used to separate these strands according to their size.

▼

The position of the repeated sequences from a single locus is identified using a radioactive probe.

3 (a) There are many different restriction enzymes. Use the information in Chapter 11 to explain why one particular sort of restriction enzyme is used to cut the DNA into pieces.

(2 marks)

   (b) A probe is a piece of DNA with the complementary base sequence to the one being sought. Explain how a radioactive probe can be used to identify the position of a particular repeated sequence of non-coding DNA.

(2 marks)

We will now go back to our original peregrine falcon family. A test was carried out on their DNA. It produced the results shown in Figure 13.18.

Figure 13.18

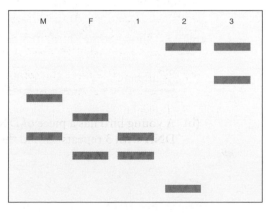

4 (a) What can you conclude from these results about the parents of the young birds? Explain your answer.

(3 marks)

   (b) Suggest one reason why care has to be taken in interpreting the results of tests like this.

(2 marks)

Examination questions

1   Table 13.2 shows the results of some blood tests carried out on a patient admitted to hospital suffering from a suspected myocardial infarction (heart attack).

Substance	Concentration in patient's blood (arbitrary units)	Range of concentration in blood of healthy individuals (arbitrary units)
Urea	5.7	2.5–6.7
Cholesterol	8.2	3.6–6.7
Lactate dehydrogenase enzyme	2263	300–600
Potassium	4.3	3.4–5.2

Table 13.2

(a)   A myocardial infarction results in damage to the muscle of the heart.

   (i)   Explain how a blood clot may cause damage to the muscle of the heart.

(2 marks)

   (ii)   Lactate dehydrogenase is an enzyme found inside healthy heart muscle cells. Suggest why the concentration of this enzyme in the blood can be used to confirm that this patient had suffered a myocardial infarction.

(2 marks)

(b)   Use the table to explain what is meant by a *risk factor*.

(2 marks)

2   Table 13.3 shows some antibiotics and the way in which they work.

Antibiotic	Method of action
Penicillin	Prevents the formation of bacterial cell walls.
Streptomycin	Distorts the shape of ribosomes in bacterial cells so that protein synthesis is stopped.
Mitomycin C	Joins together the two polynucleotide chains which make a DNA molecule with strong chemical bonds so they cannot be separated.

Table 13.3

(a) Explain why:

(i) penicillin is effective against bacteria but not against fungi;

(1 mark)

(ii) streptomycin cannot be used to treat diseases caused by viruses.

(1 mark)

(b) (i) Explain why cells treated with mitomycin C cannot synthesise proteins.

(2 marks)

(ii) Suggest why it is thought that mitomycin C might be effective in treating cancer.

(2 marks)

3 Plastic strips impregnated with certain chemicals may be dipped in urine. A change in colour indicates the presence of glucose. The test relies on the two chemical reactions shown by the equations.

$$\text{glucose + oxygen} \xrightarrow{\text{Enzyme A}} \text{gluconic acid + hydrogen peroxide}$$

$$\text{blue dye + hydrogen peroxide} \xrightarrow{\text{Enzyme B}} \text{green-brown dye + water}$$

(a) Name enzyme **A**.

(1 mark)

(b) With which of the substances shown in the equations are the plastic strips impregnated?

(1 mark)

(c) Explain why:
(i) the manufacturers recommend storing these strips in a cool place.

(1 mark)

(ii) the strips will only give a positive response to glucose. They will not give a reaction with other sugars.

(2 marks)

Index